Beyond Earth: The Search for Life's Fingerprint in Organic Matter

Nicholas

TABLE OF CONTENTS

Chapter 1

INTRODUCTION

Carbon is ubiquitous. It makes up all living things on Earth, from us to microbes to plants. It is left behind when things die. It undergoes chemical transformation to remain preserved within the rock record over time, sometimes for up to billions of years. Indeed, in some cases, such as the first half of Earth history, we only have organic matter to reconstruct what the biosphere was doing and how it was interacting with the geosphere.

Carbon is also abundant within extraterrestrial environments, and present in the samples that find their way from space to Earth (e.g., meteorites). Extraterrestrial organic molecules range in size from the very small (a few atoms) to the very large (>1000s of Daltons). For instance, polycyclic aromatic hydrocarbons, or PAHs, for short, are a class of molecules comprised of a collection of fused aromatic rings. When astronomers look out at the Milky Way galaxy and others, they see this repeated linked aromatic structure, and estimate that these ubiquitous aromatic organic molecules constitute up to 20% of all the carbon in the galaxy. The abundance of this organic matter in extraterrestrial environments alone makes it a compelling target for study. Beyond that, the parent bodies that carbonaceous meteorites and asteroids come from are also thought to be related to the planetesimals that accreted to form the Earth. Understanding the chemistry of this organic matter informs our understanding of how carbon found its way to Earth. In other words, the chemistry of organic matter formation is The Chemistry of the universe.

This book explores organic matter: where it is, what it means, and how we can study it to understand its formation and preservation in geological, planetary, and interplanetary environments (Figure 1.1). Our story begins by exploring relatively small (a few 100 Daltons), soluble molecules, such as amino acids and polycyclic aromatic hydrocarbons, which have been studied extensively through well-established analytical tools. I have worked to advance the methods available for studying these soluble compounds, as described in Chapters II-IV. Then, our story broadens, as much of the organic matter within the Earth and extraterrestrial environments is neither small nor soluble. The last two chapters of this book work

to understand the preservation of larger organic molecules within different parts of
the rock record, and the effect of large organic molecules on landscapes.

Smaller, simpler molecules

Every organic molecule is comprised of atoms, each of which has endured its
own history, thus carrying a fingerprint of where it has been and how it found
its way to its current location. Such stories can be traced using isotopes, which
follow predictable laws of chemical physics, therefore preserving history of source
and synthesis within the chemical structure of molecules. Traditionally, isotopes
of organic matter have been measured on a compound-specific or a bulk scale,
converting an entire organic molecule or group of molecules, respectively, into a
gas like CO_2 to measure its average isotopic composition. Developing methods to
analyze as many atoms as possible within each molecule expands the number of
constraints available to understand how that molecule formed, where the number of
potential dimensions available to explore are practically infinite.

Chapters II-IV contribute to the ongoing investigations of this n-dimensional
space. Chapter II describes a novel method to simultaneously measure multiple
isotopic properties for multiple compounds in a complex mixture by using gas
chromatography (GC)-Orbitrap-mass spectrometry. This method is even capable of
distinguishing near-isobars (i.e., ^{13}C and D) from one another. The direct elution
method is particularly useful when applied to the study of extraterrestrial samples,
where the range of isotopic compositions is expected to be large and the amount of
sample available for analysis is limited. Serendipitously, this method was invaluable
for performing the isotope ratio measurements presented in Chapter III. Without
this method, these measurements would not have been able to be performed at all,
let alone to an acceptable level of precision and at lower concentrations (by several
orders-of-magnitude) than those necessary for performing traditional compound-
specific isotope ratio measurements.

Chapter III characterizes the isotopic compositions of PAHs extracted from
samples returned by the Hayabusa2 sample return mission to the Ryugu asteroid.
Ryugu is thought to be similar in composition to that of a CI carbonaceous chondrite:
The meteorite class most chemically similar to the chemical composition of the Sun.
PAHs are important because of their ubiquity, as mentioned at the beginning of this
chapter, and perhaps even more so because it is not well understood how they are
formed. Some experts hypothesize that they form in extraterrestrial environments

in the same way that they form on Earth, which is through pyrolysis within high-temperature (1000K) circumstellar envelopes. However, the rate at which they would be ejected out of these envelopes into the interstellar medium (ISM) is similar to the rate at which they would be broken down within the ISM by shock wave and UV radiation. Alternatively, PAHs could be formed in low-temperature (10K) molecular clouds within the ISM, but there are limited observations to-date of this process occurring. Most notably, this chapter presents the first quantitative support for the formation of these compounds within low temperature environments, through the measurement of the doubly-^{13}C-substituted composition of these PAHs, which preserves an intramolecular record of formation temperature.

Chapter IV applies the GC-Orbitrap technology to characterize the isotopic compositions of specific atomic positions of amino acids found within the Murchison meteorite. Murchison is a well-studied carbonaceous chondrite with organic molecules present in abundances of 100s of parts per million. Prior studies of the organic contents of the Murchison meteorite have identified over 90 amino acids, including almost all of the proteinogenic ones. The carbon isotopic compositions of these amino acids support the interpretation that the compounds are exogenous to the meteorite, as the carbon isotope values range from being similar to those of proteinogenic amino acids (i.e., $\sim$-25‰) to being much more ^{13}C-enriched. Interpretations of how these compounds form in the parent body are complicated by the fact that measurements of compound-specific carbon isotopic compositions average out any intra-site isotopic variation. For instance, a prior study of α-alanine extracted from the Murchison meteorite found that the molecular average carbon isotope value of $\sim$+50‰ was driven by a highly-^{13}C-enriched amine site ($\sim$+120‰; Chimiak et al., 2021). Chimiak and others propose that the amine carbon originated from an interstellar medium-derived organic precursor, while the carboxyl and methyl sites were similar in composition to terrestrial values and therefore likely derived from carbon pools in the parent body. Thus, the position-specific carbon isotopic contents of amino acids preserve records of synthesis chemistry. In this chapter, I build on this prior work, confirming the past site-specific measurement of α-alanine by repeating it with an extract of a different chip from the same stone. I also present position-specific carbon isotope measurements for β-alanine and aspartic acid, which are two amino acids that are structurally distinct from α-alanine and therefore require different synthesis pathways. Indeed, the pattern of intramolecular isotopic composition is distinct for the three amino acids, which, when integrated with an understanding of the isotopic compositions of potential sources, informs an understanding of extraterrestrial organic synthesis networks.

The bigger stuff

I would be remiss not to mention that the soluble organic molecules described in Chapters II-IV, while important and notorious, are actually less prominent within the galaxy and the rock record than another kind of organic matter: insoluble, macromolecular carbon. Macromolecular carbon includes lignin, which is the compound that gives plants their structure, and kerogen, which constitutes the final fate of organic matter that gets preserved within rocks. Macromolecular organic matter is challenging to study, as it is heterogeneous in structure and, in the case of kerogen, it is *by definition* what is left over once everything soluble has been removed from a sample, thus rendering the insoluble residue impossible to introduce into liquid- or gas-source mass spectrometers. These features make it all the more frustrating because, as far as we know, it is the only form of carbon that is preserved within Archean rocks, and therefore the only option we have to interrogate the earliest records of life on Earth.

Chapters V and VI aim to re-address longstanding questions during pivotal moments of the rock record that relate to macromolecular carbon. Chapter V focuses on the organic carbon and kerogen found within Archean rocks. In this chapter, I present the largest compilation to-date of measurements of the carbon isotope values of organic matter found within Earth's earliest rocks. My statistical analyses of the carbon isotope contents of kerogen reveal a bimodality in the data that has never been previously observed within the literature. To interpret the lowest values within the record, I review potential reactions that can affect the isotopic composition of organic matter during the rock cycle. I incorporate these reactions into a simplified mechanistic model, which provides constraints on how much the carbon isotopic composition of organic matter can change with different amounts of post-depositional alteration and metamorphism. The model demonstrates that the most altered samples have undergone too much carbon exchange and re-crystallization to be considered primary. However, consistent with prior empirical studies, the less-altered samples may not have experienced changes in their average carbon isotopic composition by more than a few per-mille. When interpreted in the context of the data's bimodality, this result offers new motivation to re-address this long-standing problem, and poses the question: could the bimodality reflect Archean ecology?

Chapter VI provides an explanation for the irreversible and irrevocable effect that the evolution of terrestrial plants had on landscapes. Prior to the proliferation of plants on land, terrestrial landscapes were devoid of mud. Previous research has hypothesized that the subsequent 'muddying' of terrestrial landscapes was driven by plant roots lending cohesion to the soil, thus binding clay minerals in place. However, this observation has been demonstrated to be in opposition to evidence within the rock record, which demonstrates that the muddying happens prior to the proliferation of rooted plants, when the only plants that exist on Earth are small (e.g., mosses). My work proposes an alternative explanation, suggesting that the mere introduction of plant-derived organic molecules into the system could have been capable of binding clay particles together into aggregates called 'flocs,' which could in turn offer a geochemical mechanism for the rise of mud.

A final word

The stories within this book span time frames as long as the Solar System: 4.567 billion years. Within the context of this unfathomably long length of time, the lifespan of the human species, let alone this book, may seem like a speck. And yet, these stories are also testaments to how small things can affect and inspire large change: Small molecules can turn a sandy world into a big mudpit. Molecular fingerprints within a few micrograms of rock brought back from space can tell a story about where the carbon that we are made of comes from. Humans, specks that we may be, explore, sample, and analyze the Earth and the Solar System in continuing efforts to untangle its mysteries: For instance, the Perseverance rover currently traverses the surface of Mars, collecting and caching samples for their return to Earth in about a decade.

Thus, these questions are more important now than ever. It is nearly certain that organics will be identified within the samples that are returned from Mars, and likely that these organics will look less like fossils and more like macromolecular organic matter. Understanding how to analytically approach these samples is essential. Being thoughtful about the impact left behind by our visiting and sampling—either to sites on Earth, or elsewhere—is perhaps of paramount importance. I include some thoughts on this responsibility within the Appendix, which was originally published as an article in *Caltech Letters,* a student-run publication that highlights campus research and scientific perspectives. Human impact on the Earth, the rock record, and the Solar System is something that I spend a significant chunk of my own speck of time thinking about.

Ultimately, I hope this book drives you to think across different scales: spatial and temporal, big and small. If you take anything from these chapters, I hope that they empower you to see yourself beyond the speck, to learn, to discover, and, most importantly, to contribute.

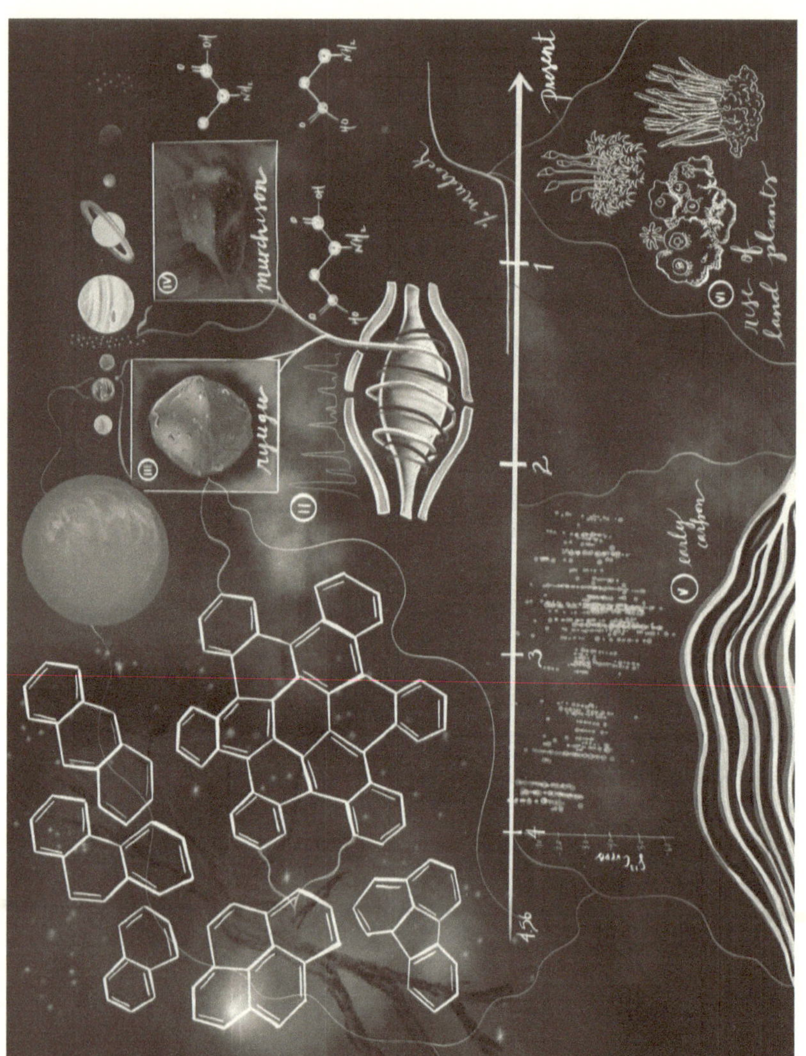

Figure 1.1: Fates of Carbon.
Caption continued on next page.

Figure 1.1: Fates of Carbon.

This figure offers a visual representation of this book as a collage. The chapters of this book include: II. Methods and limitations of stable isotope measurements via direct elution of chromatographic peaks using gas chromatography-Orbitrap. III. Polycyclic aromatic hydrocarbons in samples of Ryugu formed in the interstellar medium. IV. Position-specific carbon isotopes of Murchison amino acids elucidate extraterrestrial abiotic organic synthesis networks. V. The carbon isotopic composition of Archean kerogen and its resilience through the rock cycle. VI. Early plant organics increased global terrestrial mud deposition through enhanced flocculation.

Chapter 2

METHODS AND LIMITATIONS OF STABLE ISOTOPE MEASUREMENTS VIA DIRECT ELUTION OF CHROMATOGRAPHIC PEAKS USING GAS CHROMOTOGRAPHY-ORBITRAP MASS SPECTROMETRY

S.S. Zeichner, E.B. Wilkes, et al. (2022). "Methods and limitations of stable isotope measurements via direct elution of chromatographic peaks using gas chromotography-Orbitrap mass spectrometry". In: *International Journal of Mass Spectrometry* 477, p. 116848. DOI: 10.1016/j.ijms.2022.116848.

Sarah S. Zeichner[1], Elise B. Wilkes[1], Amy E. Hofmann[1,3], Laura Chimiak[1,2], Alex L. Sessions[1], Alexander Makarov[4], John M. Eiler[1]

Affiliations: [1] Division of Geological & Planetary Sciences, California Institute of Technology, Pasadena, CA 90025, USA; [2] Department of Geological Sciences, University of Colorado, Boulder, CO 80309, USA; [3] Jet Propulsion Laboratory, California Institute of Technology, Pasadena, CA, 91109, USA; [4]ThermoFisher Scientific, Bremen 28199, Germany

Introduction

Stable isotope ratios of natural samples are used to characterize chemical synthesis processes (e.g., to establish biogenicity), environmental conditions, and elemental budgets in natural cycles, and can be quantified at levels of natural abundance via mass spectrometry. Conventional isotope ratio measurements require an initial conversion of analyte into a simple molecular gas such as CO_2, H_2, or N_2 for isotope ratio mass spectrometry ("IRMS") using specialized magnetic-sector instruments. The resulting isotope ratio therefore represents the average for an element across an entire molecule. These techniques sacrifice the study of variation at specific molecular positions, or "position-specific" isotope ratios, and measurements of clumped or multiply-substituted isotopologues — information that could improve existing interpretations or enable new applications — in order to achieve greater precision and simplify instrumentation and methods.

Orbitrap mass spectrometry is one technique that enables direct measurements of isotope ratios of higher molecular weight compounds without requiring conversion to simple gases, thus preserving clumped and position-specific isotopic information (Chimiak et al., 2021; Eiler et al., 2017; Hofmann et al., 2020; Neubauer, Cremiere, et al., 2020; Neubauer, Sweredoski, et al., 2018). Orbitrap isotope ratio mass spectrometry has been used for tracing synthetically labeled compounds, using the natural isotope abundance profile as an aid to molecular identification (Jang, Chen, and Rabinowitz, 2018), and constraining elemental isotope ratios (Hoegg, Barinaga, et al., 2016; Hoegg, Manard, et al., 2019). These applications highlight the potential of Orbitrap mass spectrometry for geochemical studies, as the technology already serves an analogous role to simpler gas chromatography-MS (GC-MS) instruments. Further, the Orbitrap mass analyzer performs high mass resolution measurements (up to 240,000 full peak width at half maximum intensity, or FWHM, at 200 Da) capable of distinguishing near-isobaric masses (i.e., resolving isotopic substitution of ^{13}C versus ^{15}N and ^{2}H) and enabling it to function as a powerful tool for isotope ratio measurement (Chimiak et al., 2021; Eiler et al., 2017; Hofmann et al., 2020; Neubauer, Sweredoski, et al., 2018). This set of capabilities raises the possibility that the Orbitrap could be capable of continuous flow isotope ratio measurements analogous to those now made by GC-IRMS (Baczynski et al., 2018; Hayes et al., 1990; Matthews, Hayes, and Matthews, 1978), but with extended abilities to observe features of intramolecular isotopic structure.

Within an Orbitrap mass spectrometer, molecular and fragment ions form in the ion source, are accelerated through the instrument, and can be selectively filtered by mass using a quadrupole prior to injection into the Orbitrap mass analyzer (Fig. 2.2A). The mass analyzer measures the image current produced by oscillating ions (the "transient"). The relative abundances of molecular and/or fragment ions are retrieved via a fast Fourier transform of this time-dependent image current, which yields the mass-dependent frequency and relative intensity of each ion species contributing to the transient, and thus constrains the mass to charge (m/z) ratios and proportions of those species (Eiler et al., 2017; Makarov, 2000). This enables the Orbitrap to recover precise and (with adequate standardization) accurate single and multiply-substituted (i.e.,"clumped") isotopic compositions of molecules, at the per-mil-level fidelity required to study natural isotopic variations (Eiler et al., 2017). Combined with the Orbitrap's potential for molecular identification, such data can augment compound-specific isotope ratio measurements.

To optimize Orbitrap mass spectrometry for analysis of isotope ratios in natural samples, previous studies have developed techniques interfacing the Orbitrap mass analyzer with a gas chromatograph (GC; Chimiak et al., 2021; Eiler et al., 2017; Wilkes et al., 2022), as well as with liquid-medium sample introduction, electrospray ionization and secondary collisional fragmentation (e.g., Hoegg, Barinaga, et al., 2016; Mueller et al., 2021; Neubauer, Cremiere, et al., 2020). For GC Orbitrap measurements in particular, relatively high precision has been achieved by "peak capture": trapping an analyte in a reservoir configured within the GC oven and gradually releasing it to the ion source over minutes to hours in order to obtain precision for replicate analyses, in some cases approaching ~0.10‰ (Chimiak et al., 2021; Eiler et al., 2017), and thus ideal for samples with small isotopic variations. An obvious drawback to the peak capture method is that only a single peak from each gas chromatogram can be selected, so separate GC runs are required for each potential analyte.

Measuring isotope ratios by "direct elution" from the GC allows one to observe isotopic compositions of several or all compounds in a mixture as they exit the GC column, without any trapping steps. This approach sacrifices precision because the duration of Orbitrap observation is limited by the duration of analyte elution from the GC but provides the opportunity to quickly study numerous different compounds in one experiment. Further, the highest levels of precision obtained in previous studies are not necessary for all applications and could be unnecessarily time consuming when isotopic variations are large.

Here, we explore the precision and accuracy achievable using a direct elution method of isotopic analysis on Thermo Scientific™ Q Exactive Orbitrap™ mass spectrometer coupled to a Thermo Scientific™ Trace™ 1310 GC instrument. We use three examples to examine how direct elution studies can provide meaningful isotope-ratio measurements and their limits of precision: (1) para-xylene (p-xylene), (2) derivatized serine reference standards, and (3) two polycyclic aromatic hydrocarbons (PAHs; pyrene and fluoranthene) in a meteorite sample and in terrestrial standards. In each case study, we highlight how optimizing chromatographic peak shape and strategically selecting Orbitrap parameters (see glossary of terms, Table S2.1) can improve the number of observed ion counts and enable more rapid, accurate, and precise measurements of samples for compound-specific, position-specific and clumped-isotope applications. We apply these parameter selections to a numerical model of chromatographic peak elution and the Orbitrap data collection system to demonstrate the minimal effect that our proposed data processing approach has on measuring isotope ratios of a single analyte and sample-standard comparisons under typical experimental conditions, and present examples of scenarios that could produce imprecise or inaccurate values. Overall, we conclude that isotope ratio measurements using the direct elution method achieve coarser levels of precision than the highest-performance peak-trapping methods, but are still adequate for many applications and have the additional advantage of using a commercially available instrument platform.

Materials and Methods

Standards and preparatory chemistry

The following analytes were prepared for GC-Orbitrap isotope ratio measurement: para-xylene, variably ^{13}C-labeled and derivatized serine standards (SERC0 and SERC1), and polycyclic-aromatic hydrocarbons (PAHs). Full details of standard sources and preparatory chemistry are described in the supplemental material. The molecular average carbon isotopic composition of serine and PAH standards were characterized by combustion Elemental Analyzer-Isotope Ratio Mass Spectrometry (EA-IRMS) at Caltech. The mean $\delta^{13}C_{VPDB}$ values ($\pm 1\sigma$) of the variably labeled serine standards were -32.22±0.02‰ (n = 9, SERC0) and -20.35±0.02‰ (n = 6, SERC1), reflecting an enrichment of 36.8‰ in the C1 site of SERC1 relative to SERC0's. The mean $\delta^{13}C_{VPDB}$ values ($\pm 1\sigma$) for fluoranthene and pyrene were -25.68±0.66‰ (n = 3) and -25.69±0.45‰ (n = 3), respectively. In addition, a subsample of Murchison, a CM2 carbonaceous chondrite meteorite, was cleaned, powdered and dissolved in 5 mL of a 1:1 v/v solvent mixture of high-purity hexane and acetone to extract the PAHs.

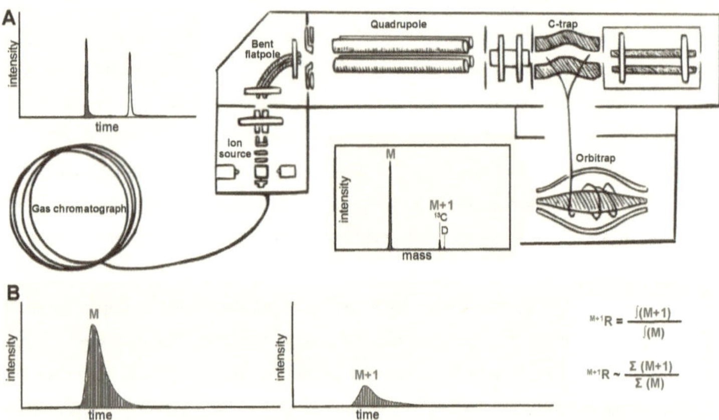

Figure 2.2: Methods.

(A) GC-Orbitrap instrumental setup. Samples are introduced via gas chromatography into the mass spectrometer, where compounds are ionized, accelerated through the bent-flatpole, and mass-separated in the quadrupole before being stored in the C-trap, and injected into the Orbitrap. Characteristic spectra of each compound are produced by the Orbitrap, which performs measurements at high resolution capable of separating ^{13}C and D substituted fragments and calculation of isotope ratios to high degrees of precision. This method probes the precision attainable within the natural GC elution time of each compound. GC, gas chromatography. (B) Isotope ratios are calculated by taking the sum of "counts," or ions, for both the substituted (M+1, in this example) and unsubstituted (M) fragments, and then taking the ratio of this sum.

Chromatography and Orbitrap mass spectrometry

Analyses were performed on a QExactive mass spectrometer with samples introduced via a Trace 1310 GC equipped with a TG-5SilMS column (30 m long, 0.25 mm ID, 0.25 μm film). Samples and standards were analyzed under closely matched experimental conditions. Analytes were injected with either a 5 μL (Hamilton #87930) or 10 μL syringe (Hamilton #80384), with target sample sizes corresponding to ~10-100 picomoles of dissolved analyte to achieve high total ion currents (TIC, 10^8 to 10^9 counts per scan), while not overloading the GC column.

Analytes were introduced via a split-splitless injector operating in splitless mode with He carrier gas (1.0-1.5 mL/min); both the flow rate and the GC oven temperature program were optimized for each analyte or mixture of interest, with the aim of creating well-separated, equant GC peaks (Table S2.2). GC effluent was transferred directly into the ion source via a heated transfer line (250°C) where analytes were ionized via electron impact (EI; Thermo Scientific Extractabrite, 70eV; Fig. 2.2A, B). Ions were extracted from the source, subjected to collisional cooling during transfer through the bent flatpole, filtered by the Advanced Quadrupole Selector[TM] (AQS) to select a particular mass window, and then passed through the automatic gain control (AGC) gate (see below) prior to storage in the C-trap — a potential-energy well generated by radio frequency and direct current potentials. Ions accumulated in the C-trap until the total charge reached a user-defined threshold (the "AGC target"), then were introduced into the Orbitrap mass analyzer as a discrete packet (Fig. 2.2A). Applications of the Orbitrap to isotope ratio analysis typically require that the mass analyzer operates under "AGC control," i.e., under conditions where the AGC target limits the number of ions admitted into the Orbitrap in each scan to a ~constant value. Within the mass analyzer, ions orbit between a central spindle-shaped electrode and two enclosing outer bell shaped electrodes, moving harmonically at frequencies proportional to m/z (Makarov, 2000). We refer to each sample analysis that requires a separate sample injection into the GC as an "acquisition"; each acquisition is comprised of "scans," where each scan is comprised of ion intensities and m/z ratios averaged by the Orbitrap over a short time interval (typically 100-300 ms, where scan length is dependent on the user-defined nominal resolution setting).

The width of the eluting chromatographic peak is central to the method, as it defines how long the compound can be observed by the Orbitrap. Because the AGC target limits the number of ions per scan, the peak width also dictates the total number of ions that can be counted. The peak width in our experiments was typically 20 seconds to 1 minute, which translated into 5-30 seconds of usable data (see "Data analysis"). This range of peak widths extends to a longer duration than is optimal for GC-MS measurements, but are necessary in order to observe the compounds of interest at a high enough ion threshold for a long enough time to achieve useful precision. The user must balance the desire for narrow peaks and good separation between co-eluting compounds with the desire to observe the compound for as many scans as possible.

We varied instrument settings — the AQS mass window, the AGC target, and the nominal Orbitrap mass resolution — to explore controls on precision, i.e., shot-noise limits and reproducibility of sample-standard comparisons of isotope ratios. Exploratory analyses were performed with a large AQS mass window to observe and characterize large parts of the fragmentation spectrum for each molecule. For better precision on measured isotope ratios of a specific molecular or fragment ion of a single compound of interest, the mass window selected by the AQS can be narrowed to increase the proportion of ions measured that come from the analyte of interest (Chimiak et al., 2021; Eiler et al., 2017).

In the majority of experiments, the AGC target was set to 2×10^5 ions, but in cases where we noticed artifacts due to ion cloud pairing, the AGC target was reduced to minimize these "space charge effects" (see "Improving precision and mitigating space charge effects in the C-trap and Orbitrap;" Eiler et al., 2017; Williams et al., 2021). In general, we chose AGC targets as high as possible without creating unacceptable artifacts on the signals of smaller peaks of interest, which we were able to discover through replicate measurements of the relative heights of peaks of interest at different AGC targets.

The resolution required to distinguish between the masses of closely-adjacent near-isobars expected to be present in the mass spectrum can be calculated as:

$$resolution = \frac{m}{\Delta m} \tag{2.1}$$

where m is the mass of an ion of interest and Δm is difference in mass necessary to separate that ion peak from an adjacent near-isobar. This required resolution is similar to the "nominal" resolution reported for the Orbitrap, which is calculated for a 200 Da fragment ion as m/dm, where m is mass and dm is the full peak width at half maximum intensity (FWHM). The choice of resolution is central to the direct elution method, because the resolution sets the time length of each scan. At higher resolutions, the Orbitrap observes the ion packet for a longer time (~300ms for 120k resolution) enabling smaller mass differences between peaks (e.g., ^{17}O and ^{13}C) to be resolved. However, these benefits come with a trade-off, as ions continue to flow through the mass spectrometer while each scan is underway. Thus, a longer time spent observing ions in the Orbitrap sacrifices more ions that reach the C-trap during the scan after the C-trap achieves the AGC target (see "Improving precision and mitigating space charge effects in the C-trap and Orbitrap;" Bills et al., 2021; Eiler et al., 2017; Hoegg, Manard, et al., 2019). To demonstrate the effects of these phenomena, a range of resolutions were explored in our experiments; a full list of experimental conditions is included in the supplemental material (Table S2.2).

Data analysis

Data from each acquisition are converted to selected-mass chromatograms (intensity versus time for a selected m/z) which can then be integrated to yield isotope ratios in a manner analogous to that used by GC-IRMS (Sessions, 2006). For this conversion, we used FT Statistic, a computer program written by ThermoFisher that extracts data from RAW files created by the proprietary Orbitrap control software. From the FT Statistic-processed files, we extracted ion intensities, peak noise, and other acquisition statistics, which we analyzed via in-house code written in Python (v3.7.6). The data analysis process is described in the subsequent paragraphs, and the processing code can be found on the Caltech data repository (Zeichner, 2021). Raw data for samples and standards were processed identically to enable comparisons between measurements, and can also be found on the Caltech data repository.

Accurate calculation of isotope ratios from direct elution experiments required careful data processing to account for time-varying intensities of fragment ions of interest. We developed a method for processing Orbitrap data that builds upon prior established methods for isotope ratio calculation in compound specific isotope ratio mass spectrometry (Ricci et al., 1994) and those established by previous studies of Orbitrap isotope ratio mass spectrometry (Cesar et al., 2019; Chimiak et al., 2021; Eiler et al., 2017; Hofmann et al., 2020; Mueller et al., 2021; Neubauer, Cremiere, et al., 2020; Neubauer, Sweredoski, et al., 2018; Wilkes et al., 2022)

Central to the direct elution method is the definition of the time frame over which to integrate peak areas. Compounds of interest can be identified by diagnostic mass spectral fragments; we define the start and end of a peak as the times immediately before and after diagnostic fragments emerge above and disappear below the instrument baseline, respectively. Stable Orbitrap scans for a given peak are achieved when the system is under "AGC control," a state where the number of ions observed by each scan is stable and meets the AGC target (roughly the product of the total ion current (TIC) and injection time (IT)). When the system operates outside of AGC control, the number of ions from scan to scan is inconsistent, which can lead to unwanted artifacts on measured isotope ratios (and often happens at the beginning of chromatographic peak elution). We observed that scans operated under AGC control when the monoisotopic peak intensity (NL score) is >10% of the maximum TIC peak intensity (peak height). We refer to this 10% threshold as the 'baseline' below, but note that it excludes some scans in which the target ions are observed at intensities greater than the true instrument baseline (i.e., signal in the absence of any analyte). We cull any scans that fall below this threshold, which provides a uniform set of criteria for defining the peak within a given window of time.

For carbon isotope ratio measurements, we perform a background subtraction by selecting a reference "background" time frame five seconds prior to the emergence of the peak and computing the average NL score of all scans within that interval for each exact mass used in the isotope ratio calculation. This background intensity is specific to each mass fragment of interest, as the background can vary for a given mass window. We subtract this background from the NL scores of each scan of our defined peaks. We did not perform background corrections for hydrogen isotope ratio (^{2}H/^{1}H) measurements, as the low natural abundance of deuterium is close to the absolute intensity of our background and use of background correction overestimated errors by eliminating a large fraction of useful ion counts.

The Fourier transform of transient data reports signal intensities (NL scores), which must be converted into the number of ions ('ion counts') to compute isotope ratios. To calculate isotope ratios for each chromatographic peak, we next converted NL scores for the scans selected after baseline culling into ion counts (Eiler et al., 2017):

$$N_{io} = S/N * C_N/z * \sqrt{R_N/R} * \sqrt{\mu} \qquad (2.2)$$

where N_{io} is the number of observed ions, S is the reported signal intensity (NL score) for the molecular or fragment ion in question, N is the noise associated with that signal, R is the formal resolution setting (defined for $m/z = 200$) used, R_N is a reference formal mass resolution at which eqn.2.2 was established, C_N is the number of charges corresponding to the noise at reference resolution R_N (4.4; a constant established by prior experiments; see Eiler et al., 2017, z is the charge per ion for the fragment of interest, and μ is the number of microscans. The number of unsubstituted and substituted ions in the scans that fell above the 10% assumed baseline were respectively summed, and then divided to calculate the isotope ratio of interest (Fig. 2.2B). We also explored calculating isotope ratios based on NL scores alone, as well as on integrated peak areas calculated by weighting ion counts for each scan by TIC intensity before summing the counts. Finally, we computed a weighted isotope ratio by taking the isotope ratio measured by each scan and weighting it by the scan's TIC intensity. All of these methods ultimately produced less accurate and precise results than the method described and applied herein, and are further described in the Supplementary Materials.

In some cases, we report isotope ratios using "delta" (δ) notation, calculated as a ratio of isotope ratios for the same molecular or fragment ion measured with identical methods and instrument conditions for the sample and reference material:

$$\delta^A X_{refmaterial} = \frac{^{A/a}R_{sample}}{^{A/a}R_{standard}} - 1 \tag{2.3}$$

where A and a are the cardinal masses of the rare and common isotopes of interest, respectively, X is the formula of the isotope-substituted element in question, and R is the observed ratio of interest. Delta values are reported with units of ‰. Depending on the compound and measurement of interest, δ values reported relative to the isotope ratios of known reference materials could be converted to one of the more widely used 'primary standard' isotope scales (e.g., VPDB; Brand et al., 2014) in cases where the ratio of interest in the standard is known on that scale (e.g., based on EA-IRMS or nuclear magnetic resonance (NMR) constraints). Systematic variation over time in the measured isotope ratio of the standard can be used to account for potential instrumental drift over the course of an analytical session, and can be incorporated into the computation of delta values, i.e., through normalizing sample ratios to the equivalent expected ratio of the standard at the time of the sample measurement; our method for taking this variation into account is described further in the Supplementary Materials.

To report the doubly-^{13}C-substituted carbon isotope ratios ($2\times^{13/12}$R) of pyrene and fluoranthene, we assume that the $2\times^{13/12}$R of our pyrene and fluoranthene standards follow a stochastic distribution corresponding to their $\delta^{13}C$ values. We use this expected distribution to compute the difference between our measured $2\times^{13/12}$R and our calculated, expected $2\times^{13/12}$R of the standards. We apply this difference to compute corrected $2\times^{13/12}$R*— which we distinguish from the R presented above with an asterisk—of our samples, which we then plot versus their $\delta^{13}C$ values relative to the stochastic fractionation line.

The reproducibility of direct elution experiments can be evaluated based on several metrics, which vary in their ability to usefully describe the precision of each measurement. First, the standard deviation of the measured isotope ratio for all scans within an acquisition ("acquisition error"; $\sigma_{acquisition}$) captures the variation of R within a single elution, but often overestimates error in R due to chromatographic variation from the start to end of an eluted peak and thus we do not explicitly report it for our experiments. Across multiple acquisitions (n), one can calculate the standard error (σ_{SE}):

$$\sigma_{SE} = \frac{\sigma}{\sqrt{n}} \tag{2.4}$$

where σ is the standard deviation of the average isotope ratios for the set of replicate acquisitions, n is the number of acquisitions. σ_{RSE} can be computed by dividing σ_{SE} by R, the measured isotope ratio. σ_{SE} and σ_{RSE} are reported in per-mil (‰). Finally, the shot noise limit represents the best-case uncertainty on a measurement of the A/a ratio and thus is a useful point of comparison with $\sigma_{acquisition}$, σ_{SE} and σ_{RSE}.

The shot noise limit is defined as:

$$\frac{\sigma_{SNL}}{R} = \sqrt{\frac{1}{\sum C_{io}} + \frac{1}{\sum c_{io}}} \tag{2.5}$$

where $\frac{\sigma_{SNL}}{R}$ is the shot noise limit on the relative standard error for a single acquisition, which is calculated based on $\sum C_{io}$ and $\sum c_{io}$ — the sums of all the counts for the substituted and unsubstituted molecular or fragment ions of interest, respectively. Note that estimates of number of ions observed in Orbitrap mass spectrometry measurements are approximate (Eiler et al., 2017) and so calculated shot noise errors should be considered rough estimates rather than precise predictions. We report errors as σ_{RSE}, unless otherwise specified.

Direct elution modeling

Prior compound-specific GC-IRMS carbon isotope analyses have documented a time offset (~60-140 milliseconds) that can occur between elution of unsubstituted and singly-^{13}C-substituted chromatographic peaks (Ricci et al., 1994), which is caused by differences in the molecular retention times of different isotopologues in chromatographic columns, driven largely by vapor pressure isotope effects (van-Hook, 1969). To evaluate the effect of this time shift on the variation in our computed isotope ratios, we modeled the time-evolution of the intensity of two hypothetical and idealized chromatographic peaks, representing two isotopologues of any given potential analyte, and parameterized by assigning different combinations of peak widths (σ), skews, and time shifts (Equations S2.1 and S2.2). For our model runs, we held the AGC target, nominal resolution and maximum injection time constant (2×10^5, 120k, and 3000 ms, respectively), which together determine how many ions are observed in a single scan and how long the Orbitrap observes them for.

Peak parameters and data processing decisions (e.g., thresholds on the rise of peak intensity above instrument baselines, which are used to cull chromatograms and define integrated peak areas) were then varied to represent a range of potential scenarios we could encounter in our instrumental system and to illustrate the impact of different data processing choices on the reported isotope ratio. We computed the difference between the "known" ratio of the unsubstituted and isotope-substituted forms of the idealized eluting compound and the ratio returned by our data analysis method:

$$error = \frac{(unculled - culled)}{unculled} * 1000 \tag{2.6}$$

which we multiply by 1000 to report in ‰ units. Here, "culled" refers to the scans retained after taking into account the sensitivity of the measurement (i.e., how many of the ions introduced into the instrument are observed by the Orbitrap; Eiler et al., 2017) and the post-processing culling of scans, typically rejecting those that fall below the 10% maximum NL score of a given chromatographic peak. We evaluate error for both the absolute isotope ratio measurement of a single modeled analyte $^{A/a}R_{sample}$ over a range of varied parameters, as well as the error for an idealized sample-standard-comparison $\frac{^{A/a}R_{sample}}{^{A/a}R_{refmaterial}}$ to evaluate the effects of the varying relative isotope abundances between sample and standard (a substituted-sample/a substituted-standard = 0.001, 0.01, 0.1, 1) as well as differences in absolute measured concentrations. The full details of the model can be found in "Chromatographic peak model details" of the Supplementary materials.

Results

Para-xylene

Electron-impact ionization of p-xylene produces both a strong molecular ion (m/z = 106.168 Da) and a second major tropylium fragment ion formed by loss of the methyl group (91.144 Da; these two are subsequently referred to as the 106 and 91 ions, respectively; Fig. 2.3A). Para-xylene peaks typically eluted over periods lasting ~20 seconds above baseline, but the scans that were included in the signals we integrated for isotope ratio measurement typically sampled a time frame of six seconds due to the exclusion of any scans less than 10% of the maximum peak height; those low intensity scans made up the extended 'tail' of the peak that constituted most of its observed duration. For each experiment, we present σ_{RSE} for multiple acquisitions, with isotope ratios computed as described in Section 2.3 (note that n varies for different experiments).

For p-xylene fragments measured with a large AQS mass window (80-120 Da), we achieved the best precision at 15k nominal mass resolution with an AGC target of $2{\times}10^5$; σ_{RSE} were 1.3‰ and 2.1‰ for the $^{13/12}R_{p-xylene}$ std of 91 and 106 fragments, respectively (n=5; Fig. 2.3B). However, at this resolution ^{13}C and ^{2}H are not resolved from one another for either fragment, which may be important for some applications. By increasing the resolution to 60k, ^{13}C and ^{2}H were resolved from one another but the relative intensity of the ^{2}H mass spectral peak was still suppressed from the expected ratio (~100:1 ratio in terrestrial samples) such that the $^{2/1}R_{p-xylene}$ std would not be reliable. Under this nominal resolution, with an AQS of 80-120 Da and AGC target of $5{\times}10^6$, we achieved σ_{RSE} of 2.2‰ and 2.4‰ for the $^{13/12}R_{p-xylene}$ std of 91 and 106 fragments, respectively (n=5; Fig. 2.3C). At the highest resolution (120k), AQS mass window of 80-120 Da, and $2{\times}10^5$ AGC target, we were achieved σ_{RSE} of 3.1‰ and 3.4‰ for the 91 and 106 fragments, respectively (n=10; Fig. 2.3D). At this resolution, despite being able to fully resolve ^{13}C and ^{2}H, the large mass window made it so that not enough ^{2}H atoms could be observed for the limited number of scans of the elution. For these cases, we explored measuring $^{2/1}R_{p-xylene}$ std with a smaller mass window.

By narrowing the AQS window to a 2 Da range (105.5 to 107.5 Da), to focus on only the unsubstituted and substituted molecular fragment ions, we were able to characterize both the $^{13/12}R_{p-xylene}$ std and the hydrogen isotope ratio ($^{2/1}R_{p-xylene}$ std) to σ_{RSE} of 5.4‰ and 107‰, respectively (n=5; Fig. 2.3E). Thus, the 2 Da AQS window facilitates measurement of the hydrogen isotope ratio, but to low precision.

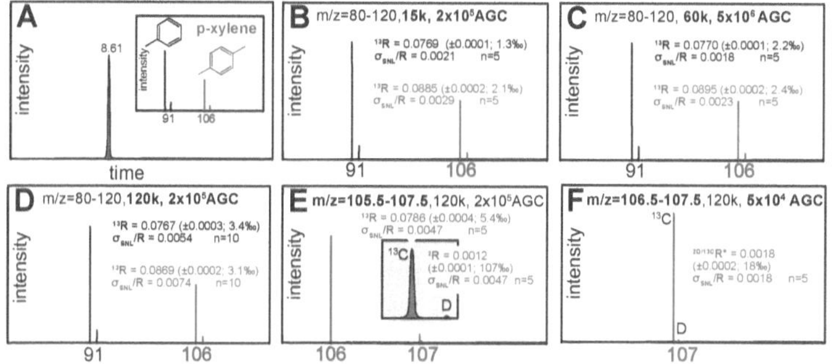

Figure 2.3: Para-xylene results.

(A) The p-xylene mass spectrum has a bright molecular ion (M=106Da, red) and another strong fragment that is the molecular ion (M) with loss of methyl group (91Da, black). (B-F) Measurements of p-xylene under a variety of experimental conditions with reported average isotope ratios, R ($\pm\sigma_{SE}$, σ_{RSE} (‰)), which are calculated based on number of counts and compared to predicted instrumental shot noise limits ($\frac{\sigma_{SNL}}{R}$). (B) Large mass window (80-120) at low resolution (15k), and moderate AGC target (2×10^5) enables experimental reproducibility of σ_{RSE} =1.3‰ at best. (C) When p-xylene is measured with increased resolution—60k—and an AGC target of 5×10^6, or (D) 120k and an AGC target of 2×10^5, precision decreases, but the baseline separation of ^{13}C and ^2H may be necessary and advantageous for some studies. (E) At high resolutions and small mass windows, ^2H and ^{13}C fragments can be resolved in direct injection method. The carbon isotope ratio ($^{13/12}R$) for M is measurable at experimental reproducibility comparable to shot noise for a mass window of M±2; (F) experimental reproducibility of the substituted hydrogen mass spectral peak (^2H) can be improved when the mass window is decreased to focus exclusively on the M+1 fragment under lower AGC control.

By narrowing the AQS window to a 1 Da range ("M+1 experiment;" 106.5 to 107.5 Da), and lowering the AGC target to 5×10^4, we were able to better constrain the reproducibility of the measured ^2H-substituted fragment ion abundance relative to ^{13}C-substituted fragment ion, represented by a σ_{RSE} of 18‰ (n=5). This ratio would have to be paired with a one of our other $^{13/12}R_{p-xylene}$ std measurements to provide a useful isotope ratio measurement. In all cases, the measurement precision (σ_{RSE}) across replicate acquisitions closely approached the estimated shot noise limits ($\frac{\sigma_{RSE}}{R}$) of the instrument based on calculated number of ions (Fig. 2.3B-E).

Together, the measurements of the m/z 91 and 106 fragments can constrain the average carbon isotopic composition of the methyl positions in p-xylene, like a previous study that used fragment measurements to compute the position-specific isotopic composition of carbon sites in ethyl toluenes (Cesar et al., 2019). Likewise, one could extend the M+1 experiment to constrain position-specific hydrogen isotopic composition of the methyl positions.

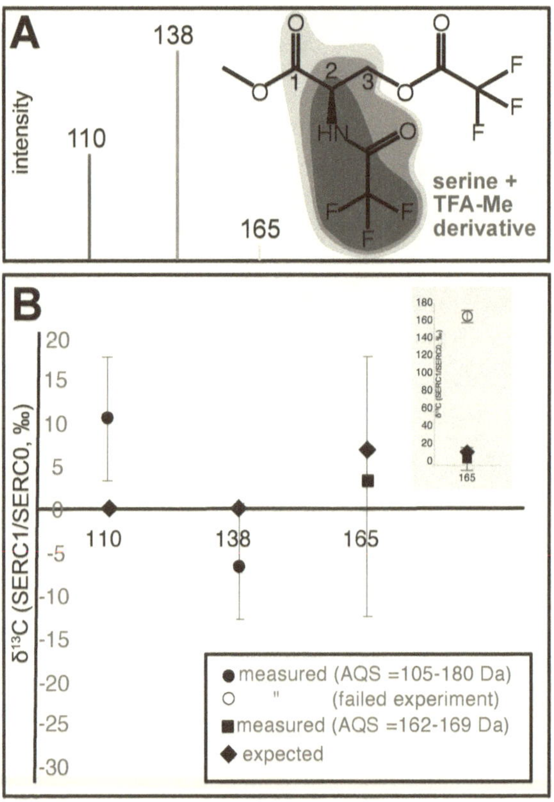

Figure 2.4: Serine results.

Caption continued on next page.

Figure 2.4: Serine results.

(A) Serine, depicted with a TFA-methyl ester derivative, produces a more complex fragmentation pattern, whose smaller fragments are measured reproducibly by direct injection, with $\sigma_{RSE} \sim \frac{\sigma_{SNL}}{R}$. Note that the 110Da fragment is a recombination product and not a direct cleavage from the original derivative. (B) Measured serine fragment isotope ratios ($\pm \sigma_{SE}$) compared to expected serine fragment isotope ratios, constrained by offline measurement by EA-IRMS. For wide AQS mass window experiments, the measured isotope ratios of the 110 and 138 Da fragments were consistent within 2 and $1\sigma_{SE}$ of expected ratios, respectively; the 165 Da fragment isotope ratio measurement became consistent within $1\sigma_{SE}$ of the expected ratio when the AQS window was narrowed to 7 Da. (inset) Comparison of measurement of 165 Da fragment with large (open black circle) versus small (filled black square) AQS mass window demonstrates the inability of large AQS mass window measurement to provide accurate isotope ratio measurements for minor mass spectral peaks (e.g., for fragment ions that form inefficiently during ionization) via the direct elution method.

Serine

Serine standards with known differences in molecular-average and position-specific carbon isotope ratios offer a well constrained test of the direct elution method's performance for fragment ions containing different subsets of atoms from the parent molecule. This test holds particular relevance for position-specific isotope applications using the direct elution method, and may also be representative of performance characteristics for compounds that produce complex fragmentation spectra, low-intensity fragment ions, and no detectable molecular ion (Fig. 2.4A).

Our case study included fragment ions of a derivatized serine standard having natural isotope abundances ('SERC0') and another to which a site-specific ^{13}C label had been added at the carboxyl carbon (C-1) site ('SERC1'). Expected site-specific carbon isotope ratios relative to the SERC0 working standard were calculated based on our knowledge that N,O-bis(trifluoroacetyl) serine methyl ester forms three major fragments during ionization. The 110, 138 and 165 Da fragments sample different combinations of carbon atoms originating from structurally distinct sites of the parent serine molecule and the derivative groups (Fig. 2.4A; confirmed by label addition studies and comparison of exact masses with spectral prediction software). The 165 Da fragment is the only characterized fragment ion containing C-1 from the parent serine molecule (i.e., the carboxyl moiety), together with 4 additional carbons that contribute to the 165 Da fragment ion. Independent constraints indicate the C-1 position of SERC1 is 36.8‰ higher in $\delta^{13}C$ than that position in SERC0, and all four of the other carbons contained in the 165 Da fragment ion of SERC1 and SERC0 should have identical $\delta^{13}C$ values to one another. Thus, the expected value of $\delta^{13}C_{SERC0}$ of SERC1 for the 165 Da fragment is 7.4‰. The 110 Da and 138 Da fragments do not include C-1, and so the expectation is that $\delta^{13}C_{138Da,SERC0} = \delta^{13}C_{110Da,SERC0} = 0‰$.

Exploratory measurements of derivatized serine were performed with a 105-180 Da AQS window, 120k nominal resolution, and 2×10^5 AGC target. Focused measurements of the least abundant fragment ion (monoisotopic mass 165.0037 Da) were performed with an AQS mass range of 162 to 169 Da to explore achievable accuracy and precision for mass spectral peaks with lower relative intensities. In all measurements, serine derivative eluted for ~1 minute, but the scans used for the calculation of the isotope ratio comprised just 15 seconds due to a broad peak shape and culling rule. For the exploratory wide AQS mass window measurements, we measured $\delta^{13}C_{SERC0} \pm \sigma_{RSE}$ values of -7.4 ± 8.1‰ and 11.8 ± 7.8‰ for the 138 and 110 Da fragment ions, respectively ($n_{SERC0} = 2$, $n_{SERC1} = 3$; Fig. 2.4B). Isotope ratio measurements performed with a large AQS mass window were consistent with our expectation that $\delta^{13}C_{SERC0}$ would equal 0‰ to within 1SE and 2SE, respectively (Fig. 2.4B). Measurements of the 165 Da fragment ion with the larger, exploratory AQS window did not agree with expectations within $2\sigma_{RSE}$ (inset Fig. 2.4B). However, the accuracy improved, matching expectations within $1\sigma_{RSE}$, by using a narrower mass window from 162 to 169 Da ($\delta^{13}C_{SERC0} \pm \sigma_{RSE} = 3.2 \pm 13‰$; inset Fig. 2.4B, Table S2.2).

Polycyclic aromatic hydrocarbons

Polycyclic aromatic hydrocarbons (PAHs) fluoranthene and pyrene represent an application of the direct elution method to the analysis of organics extracted from natural samples. Pyrene and fluoranthene were isolated from the Murchison meteorite, and the proportions of singly-and doubly-^{13}C-substituted forms of their molecular ion peaks were measured and standardized by comparison with terrestrial materials (Fig. 2.5). Measurements were performed with an AQS window of 198 to 206 Da, 2×10^5 AGC target, and nominal resolution of 180k. Pyrene and fluoranthene peaks eluted for ~40 seconds, but the selected scans for isotope ratio analysis represented a time frame of 35 seconds.

Pyrene and fluoranthene $\delta^{13}C_{VPDB} \pm \sigma_{RSE}$ values were -11.8 ± 2.4‰ and -3.9 ± 2.3‰, respectively (Fig. 2.5C&D; Table S2.3). Our direct-elution Orbi-

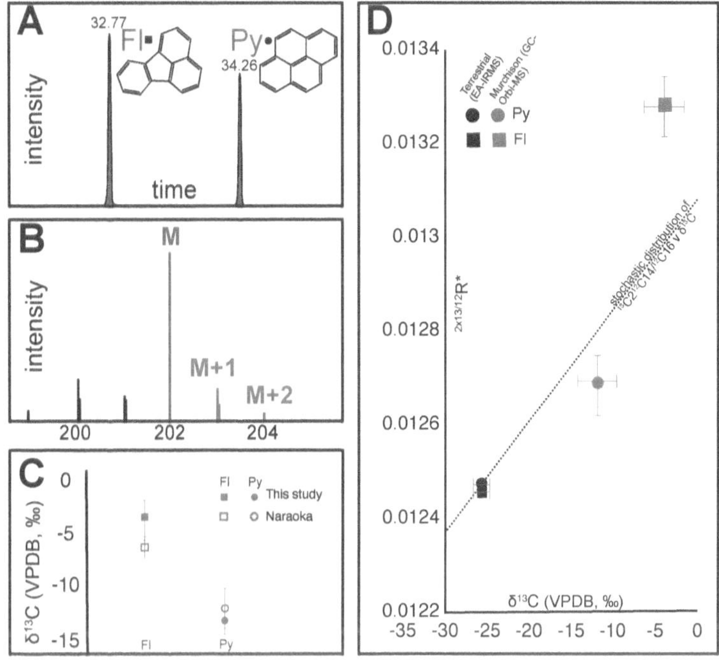

Figure 2.5: PAH results.

(A) GC elution of two polycyclic aromatic hydrocarbon (PAH) isomers, Fluoranthene (Fl) and Pyrene (Py). Fl and Py share (B) same mass spectrum, which is dominated by the presence of a strong molecular ion (M). (C) Application of direct injection method to measurement of soluble PAHs extracted from the Murchison meteorite. Error bars represent $1\sigma_{RSE}$. Measured Fl and Py $\delta^{13}C$ are consistent within $1\sigma_{RSE}$ with previous measurements; Naraoka, Shimoyama, and Harada, 2000. (D) Singly- and doubly- ^{13}C-substituted pyrene (blue circle) and fluoranthene (blue square) stable isotope ratios from Murchison measured via GC-Orbitrap-MS. They are plotted compared to terrestrial standards (black circle and square); $\delta^{13}C_{VPDB}$ values were measured via EA-IRMS, $2\times^{13/12}R*$ are corrected based on the assumption that our standards are stochastic and the offset is applied as a correction to our measured sample $2\times^{13/12}R$ values). For reference, we plot the stochastic distribution of $2\times^{13/12}R$ versus $\delta^{13}C_{VPDB}$ (dotted line). $2\times^{13/12}R$ of pyrene is lower than the relative to the expected distribution and fluoranthene's is higher.

trap measurements are consistent within $2\sigma_{SE}$ with previously reported GC-IRMS compound-specific isotope analyses of these compounds from the Murchison meteorite (-13.1 ± 1.3‰ and -5.9 ± 1.1‰ for pyrene and fluoranthene, respectively, Fig. 2.5C, Table S2.3; Naraoka, Shimoyama, and Harada, 2000). Our corrected doubly-substituted pyrene and fluoranthene carbon isotope ratios ($2\times^{13/12}R\pm\sigma_{SE}$) were 0.0121 ± 0.0001($\sigma_{RSE} = 9.5$‰) and 0.0127 ± 0.0001($\sigma_{RSE} = 8.6$‰), respectively (n = 2), which are respectively depleted and enriched within error relative to the expected stochastic distribution at their respective $\delta^{13}C$ values (Fig. 2.5D, Table S2.3).

Model results for effects of varying time shifts between chromatographic peaks

In our experiments presented above, we observed "time shifts" between the elution curves of unsubstituted and substituted ions, represented by variations in isotope ratios calculated for each scan as a function of elution time (Fig. 2.6). Experimental reproducibility of the measurements presented above generally follows expected shot-noise limits, which suggests that the method and data processing procedures we present do not introduce substantial biases in isotope ratios due to variable sampling of 'time shift' effects. Nevertheless, it is imaginable that such biases could occur for different analytes and under different conditions (Matucha et al., 1991).

To evaluate the possibility of peak shift effects on isotope ratio measurements, we applied a Monte Carlo approach to our chromatographic peak model. We chose common experimental parameters for the modeled chromatographic peaks (unsubstituted peak maximum NL score = 1×10^9 with 1% relative abundance of the substituted isotopologue; Fig. 2.6A). Using these initial peak parameters, we varied the peak shape and data analysis parameters by iterating through 3600 different combinations of "time shift" (i.e., difference in time between eluting substituted and unsubstituted fragment ions), peak width, peak skew, and peak baseline threshold (i.e., intensity threshold above baseline at which peaks were integrated), using the mathematical model of peak elution (Eqns. S2.1 and S2.2). Our model iterations produced an average error between "actual" and "computed" absolute isotope ratios of 2.9‰ with a standard deviation of 3.7‰. In all of our iterations, we observed that three specific parameters — the chromatographic peak time offset, GC peak width (σ), and the peak baseline culling threshold — together produced the most extreme errors (Fig. 2.6B).

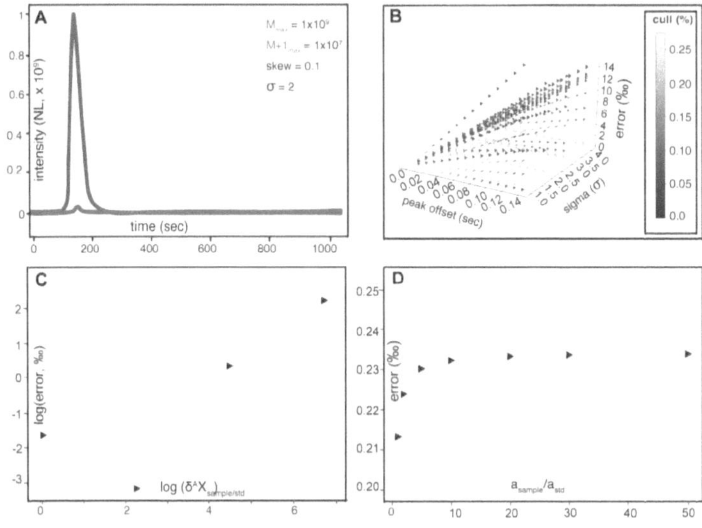

Figure 2.6: Model results.

(A) Schematic of one model run, for a $\frac{M+1}{M} = 0.01$, a skew of 0.1 and a peak width with $4\sigma = 8$ seconds. (B) Error (‰) versus peak offset (seconds) and peak width (σ) produced by 3600 model iterations, color coded by the % of maximum peak height culled. (C) Log(error, ‰) versus log($\delta^A X_{\frac{sample}{standard}}$) from 1-1000‰ difference. Errors do not exceed σ_{SNL}/R until sample-standard differences are on the order of 100‰. (D) Error (‰) versus $a_{sample}/a_{standard}$, which represents the relatively minimal effect of concentration variation between sample and standard on measured isotope ratio errors within σ_{SNL}/R.

This modeling exercise examined chromatographic peak widths ranging from 30 to 200 seconds in total elution time, which translated to 10 to 100 second periods of peak integration, after excluding usable scans having intensities that fell below the specified baseline. We apply this model to study how systematic errors arising from biased sampling of chromatographic retention time shifts may translate to errors in sample-standard comparisons, assuming 1, 10, 100, and 1000‰ contrasts in isotope ratio between model sample and model standard. We held the other parameters constant at representative values (unsubstituted peak NL score = 1×10^9 with 1% relative abundance of the substituted isotopologue, skew = 0.1, $\sigma = 2$, time offset= 100 ms), and defined a constant background for both the sample and standard three orders of magnitude lower in intensity than the maximum for the standard peak

$(a_{std} \times 10^{-3})$. As sample-standard differences in isotope ratio increased from 1 to 1000‰, the errors also increased, exceeding σ_{SNL}/R when the sample-standard isotope differences were on the order of 100‰ and above. Differences in absolute concentration (a) of sample relative to standard ($a_{sample/std}=1, 1.1, 1.2, 1.5, 2, 5, 10$) also produced increasing errors in the sample-standard differences in isotope ratios, but they are always less than 1‰ (Fig. 2.6D). Overall, when sample and standard model conditions were matched, even with parameter choices that would create variable peak shapes causing inaccurate absolute isotope ratios, we found computed sample-standard isotope ratio comparisons to be consistent with the expected isotope ratios between model sample and standard.

Discussion

The direct elution method on the GC-Orbitrap enables high mass resolution isotope ratio measurements that are precise to per-mil levels and accurate when standardized. Precision and accuracy of isotope measurements using the direct elution method can be improved by optimizing the properties of peaks eluted from the GC and the settings of the Orbitrap mass spectrometer, consistent with findings from past studies (Hoegg, Marcus, et al., 2018; Neubauer, Cremiere, et al., 2020; Su, Lu, and Rabinowitz, 2017; Uechi and Dunbar, 1992). The following paragraphs discuss effects that can be encountered in a direct elution measurement and recommend best practices.

Chromatography

Our model demonstrates how chromatographic effects and data culling methods could lead to artifacts for both absolute isotope ratios and sample-standard isotope ratio comparisons. The largest errors for absolute isotope ratios (>3‰) occurred in model iterations with some combination of large peak widths ($4\sigma > 12$ sec), large time shift between unsubstituted and substituted chromatographic peaks (>100 msec), and high baseline culling threshold (i.e., peak tails were "clipped"); the worst case scenario of this method occurs when one of these conditions are very large/high, or two or more of these parameters at relatively large/high. This result is consistent with previous research that demonstrated that inconsistent integration of substituted and unsubstituted chromatographic peaks due to large chromatographic time shifts can produce inaccuracies in measured isotope ratios (Matucha et al., 1991). For moderate parameter choices more representative of typical chromatographic conditions and culling decisions for our experiments, this model demonstrates that our method does not impart meaningful artifacts on measurement accuracy or precision, within other sources of error (which are dominated by shot noise errors). Further, the largest errors in absolute isotope ratio measurements are minimal when incorporated into a sample-standard comparison where the sample and standard are analyzed under identical experimental conditions. It is important to note however that careful sample-standard matching with respect to both isotopic composition and rate of delivery to the ion source is important for this method to be effective; large differences in isotopic composition and absolute intensity between sample and standard can also introduce errors in measured sample-standard differences in isotope ratio (Fig. 2.6C, D).

The model results also highlight an optimal chromatographic peak width for this method: 5-40 seconds of usable measurement time. Peaks that are too narrow are affected by the time shift and produce dramatic errors in isotope ratio measurement. Wider chromatographic peaks can improve precision by increasing length of time that the NL score is above the chosen culling threshold and thus the length of time that the Orbitrap can observe the analyte. However, peaks that are too broad may prevent one from adequately separating multiple compounds in a complex mixture. Tailing peaks may present similar separation issues, and may not even provide the benefit of additional scans on the tail to observe the compound of interest (i.e., if the NL score of the tail falls below the 10% baseline and thus gets culled). Too much analyte injected for an acquisition may cause chromatographic peak fronting and column overloading. All our experiments were performed within the optimal peak shape regime, which is reflected in our experimental errors that are comparable to shot noise errors.

Improving precision and mitigating space charge effects in the C-trap and Orbitrap

There is an optimal number of ions per scan that can be observed due to the behavior of the AGC target and space charge effects . Operating the mass analyzer outside of "AGC control" can cause variations in the number of ions observed from scan-to-scan, and lead to inaccurate and imprecise measurements of isotope ratios. This is partially driven by space charge effects, a phenomenon by which ions affect the movement and trajectory of one another, disturbing them from the persistent, harmonic orbits required for Fourier-transform mass spectrometry (Eiler et al., 2017; Gordon and Muddiman, 2001; Hoegg, Marcus, et al., 2018; Hofmann et al., 2020; Neubauer, Sweredoski, et al., 2018; Su, Lu, and Rabinowitz, 2017; Uechi and Dunbar, 1992). Over the length of measurement of the transient, Coulombic interactions between ions cause them to oscillate, decay, and be ejected from the mass analyzer, often at different rates for species of different masses. This effect is more pronounced in complex mixtures (Hofmann et al., 2020), samples with high levels of contaminants similar in mass to the fragment ions of interest, or measurements of less abundant fragment ions (Eiler et al., 2017; Gordon and Muddiman, 2001; Hoegg, Marcus, et al., 2018; Su, Lu, and Rabinowitz, 2017; Uechi and Dunbar, 1992).

The precision of isotope ratio measurements can be improved by lowering the nominal mass resolution, because it reduces the length of time the ion packet is observed for each scan and thus enables overall observation of more scans, and thus more counts per acquisition (Fig. 2.3B versus Fig. 2.3C). However, in some cases, reducing resolution too far may sacrifice the ability to mass resolve fragments of interests from near-isobars (e.g., ^{2}H coalescing with ^{13}C at 15k resolution). The precision on the measured isotope ratio can also be improved by increasing the number of acquisitions of a single analyte, which will decrease σ_{RSE}.

Overall, the direct elution method works best when the fragment ion of interest is the dominant ion in the mass window. In the para-xylene experiments, we demonstrate that the precision of $^{2/1}$R measurements and the influence of space charge effects can be improved when the mass window is narrowed and the ion load in the C-trap and Orbitrap is reduced such that the relative abundance of ^{2}H with respect to the total population of ions is much higher (Fig. 2.3F). Similarly, in our serine case study, we achieved the best precisions for measurements of larger fragment ions, likely because low-intensity ions may be compromised when analyzed along with more intense ions due to space charge effects and coalescence (Eiler et al., 2017; Gordon and Muddiman, 2001; Hofmann et al., 2020). We observed that this effect was mitigated when smaller fragments were measured with a narrowed mass window (Fig. 2.4). Improving measurement precision for each fragment becomes even more important if the study requires measurements of position-specific isotope values, since these fragment-specific measurement errors often compound to much larger errors in calculated position-specific isotope ratios (Chimiak et al., 2021; Neubauer, Sweredoski, et al., 2018).

In some cases, isolation of masses of interest with the AQS may not be possible, thus forcing a case where background ions impose space charge effects on target analyte ions, potentially changing their measured isotope ratios (Hofmann et al.,

2020). This effect has only been shown to influence isotope ratios when non-analyte ions make up more than 20% of the total ion current (TIC; Hofmann et al., 2020), and has been demonstrated to have a smaller effect on the measured isotope ratio at lower resolutions (i.e., 60k). The presence of extraneous ions can be mitigated by increasing the amount of measured analyte compared to the amount of contaminant, improving chromatographic separation between species of interest and co-eluting analytes, preparatory offline separation/purification of analytes, or matrix matching samples and standards that have been intentionally contaminated such that standards and samples are affected by contaminant in the approximately the same amount (Hofmann et al., 2020).

Potential applications of the direct elution method

The measurements of the fragment ions of serine by the direct elution method provide an example of using direct elution stable isotope ratio measurements via GC-Orbitrap to constrain position-specific isotopic variations in biologically relevant compounds with complex fragmentation patterns. Overall, the precisions on the fragment ion isotope ratios of serine ($\sim$8-13‰, $1\sigma_{RSE}$) were substantially worse than typical precisions achieved for molecular-average $\delta^{13}C$ measurements of derivatized amino acids by GC-IRMS ($\sim$0.1-1‰, Silverman et al., 2021) or for fragments of amino acids characterized by GC-Orbitrap using non-traditional sample introduction methods ($\leq$1‰ with 'peak trapping'; Eiler et al., 2017; Wilkes et al., 2022. Nonetheless, the direct elution method provided isotopic information about three fragments of derivatized serine that are inaccessible by conventional isotope ratio mass spectrometry, providing an avenue to position-specific isotopic information in carefully-selected contexts. However, these improvements may not suffice for the measurement of natural abundance isotope ratios in compounds with complex fragmentation spectra.

Thus, for compounds with fragmentation characteristics similar to serine, the direct elution method is best reserved for isotopically labeled samples or extraterrestrial materials where isotopic enrichments significantly exceed the measurement precision. For instance, this method could be applied to amino acid mixtures, which could be quickly surveyed for the presence of large (10's of per-mil) site-specific variations. Isotopic labels of this magnitude are often used to trace compounds and their constituent atoms through biosynthetic pathways (e.g., for drug metabolism studies; Abramson, 2001; Thevis, Krug, and Schänzer, 2006); isotope ratio measurements via direct elution to the GC-Orbitrap offer a novel technique that could detect lower levels of isotopic labels which in turn may be cheaper and safer than existing methods. Within natural samples, organic compounds extracted from carbonaceous chondrites exhibit extreme isotopic enrichments compared to terrestrial samples (measured differences in molecular average $\delta^{13}C$ values up to $\sim$50‰; $\delta^{15}N$ up to $\sim$80‰; δ^2H up to $\sim$1000's of ‰ (Glavin et al., 2018); site-specific $\delta^{13}C$ values up to $\sim$140‰; (Chimiak et al., 2021)). For samples with subtler intramolecular isotopic contrasts, longer sample introduction methods (e.g., peak trapping) remain the more appropriate choice.

The measurements of PAHs extracted from the Murchison meteorite provide a compelling example for the application of the direct elution technique to natural samples. In particular, fluoranthene and pyrene are structural isomers with the same molecular weight (202 Da), are the most abundant PAHs found on the Murchison meteorite ($\sim$1μg/g), and are known to be $\sim$10-20‰ higher in $\delta^{13}C_{VPDB}$ than terrestrial PAHs — a strong signal that makes them ideal targets for studies using the methods presented here. Enrichments of the singly-^{13}C substituted molecular ions of fluoranthene and pyrene from Murchison observed in this study are consistent, within error, with previous CSIA measurements that demonstrate enrichments compared to terrestrial standards (Figure 2.5C; Naraoka, Shimoyama, and Harada, 2000). Specifically, the data presented here (Figure 2.5C; Table 2.3) suggest Murchison's fluoranthene and pyrene have pronounced (10's of per-mil) enrichment and depletion, respectively, in their doubly-^{13}C-substituted isotopic compositions relative to a stochastic distribution. We are not aware of any prior published measurements of abundances of doubly-^{13}C-substituted PAH's from this (or any other) sample, but we note that common physical and chemical fractionations at Earth-surface conditions generally lead to only subtle ($\sim$1 ‰) variations in doubly-^{13}C-substituted forms of organic molecules (e.g., Clog et al., 2018). While a full interpretation is beyond the scope of this methodological study, these measurements suggest the potential of this method to fuel significant new discoveries regarding the origins of hydrocarbons in primitive meteorites.

Together, the serine and PAH case studies suggest that one promising future application for the direct elution method lies in its translatability to study of extraterrestrial organics, specifically, sample introduction in Orbitrap-based mass spectrometers being developed for spaceflight (e.g., Arevalo et al., 2018). Gas chromatograph-based mass spectrometers (GC-MS) have long been used as instruments for in-situ separation and analysis of organic species in planetary environments

(Anderson et al., 1972; Eigenbrode et al., 2018; Freissinet et al., 2015; Mahaffy et al., 2012; Niemann et al., 2010); the GC-Orbitrap presents a compelling future opportunity for this work. In the quest to find biosignatures or 'prebiotic' molecules on ocean worlds like Enceladus, Europa, or Titan, the direct elution method could advance our understanding of alien worlds like few other instruments and methods currently in development.

Conclusion

This study demonstrates an optimized method for quick isotopic analysis of complex mixtures via Orbitrap mass spectrometry. We present results from experiments and a chromatographic model to highlight the ability of this method to produce accurate and precise isotope ratios for singly- and doubly-substituted isotope anomalies, including position-specific isotopic properties, at precisions sufficient for studies of samples that bear high-amplitude isotopic signatures. In all case studies, our errors approach shot noise limits which demonstrate that we are achieving the highest precision possible given the number of ions that the Orbitrap can measure within the time frame of elution. This level of precision may be sufficient for some cases (i.e., extraterrestrial applications, labeling studies), while other applications may prefer other Orbitrap or MS methods. There may be potential for further development of direct elution method, with additional focused studies of lower abundance fragments and exploration of the direct elution method to measure isotopic abundances for compounds within complex sample matrices. We envision that, designed correctly, direct elution experiments have the potential to drive novel environmental and in-situ analyses on Earth, enhance the study of commercially enriched isotopically labelled materials, and facilitate future projects to Mars, Enceladus, and other extraterrestrial bodies.

Funding: Funding for SSZ was provided by NSF GRFP and a NASA Emerging Worlds grant to AEH (grant number 18-EW18_2-0084). This work was supported in part by grants from NASA Astrobiology Institute (grant number $80NSSC18M094$ to ALS and JME), the Agouron Institute (grant number AI-F-GB54.19.2 to EBW), and Caltech's Center for Environmental Microbial Interactions (CEMI, to EBW, ALS, and JME). A portion of this work was performed at the Jet Propulsion Laboratory, which is operated by the California Institute of Technology under contract with the National Aeronautics and Space Administration ($80NM0018D004$); AEH was supported by Emerging Worlds grant (grant number $18 - EW18\_2 - 0084$).

Data and materials availability: All RAW data and code is available within online repositories (Zeichner, 2021; Zeichner et al., 2021), and cited in the main text or the Supplementary Materials.

References

Abramson, F.P. (2001). "The use of stable isotopes in drug metabolism studies". In: *Seminars in Perinatology* 25, pp. 133–138. DOI: 10.1053/sper.2001.24568.

Anderson, D.M. et al. (1972). "Mass spectrometric analysis of organic compounds, water and volatile constituents in the atmosphere and surface of Mars: The Viking Mars Lander". In: *Icarus* 16, pp. 111–138. DOI: 10.1016/0019-1035(72)90140-6.

Aponte, J.C., J.P. Dworkin, and J.E. Elsila (2015). "Indigenous aliphatic amines in the aqueously altered Orgueil meteorite". In: *Meteorit Planet Sci* 50, pp. 1733–1749.

Arevalo, R. et al. (2018). "An Orbitrap-based laser desorption/ablation mass spectrometer designed for spaceflight". In: *Rapid Communications in Mass Spectrometry* 32.

Baczynski, A.A. et al. (2018). "Picomolar-scale compound-specific isotope analyses". In: *Rapid Communications in Mass Spectrometry* 32.

Bills, J.R. et al. (2021). "Improved Uranium Isotope Ratio Analysis in Liquid Sampling Atmospheric Pressure Glow Discharge/Orbitrap FTMS Coupling through the Use of an External Data Acquisition System". In: *Journal of the American Society of Mass Spectrometry* 32. DOI: 10.1021/jasms.1c00051.

Brand, W.A. et al. (2014). "Assessment of international reference materials for isotope-ratio analysis (IUPAC technical report)". In: *Pure and Applied Chemistry* 86, pp. 425–467.

Cesar, J. et al. (2019). "Isotope heterogeneity in ethyltoluenes from Australian condensates, and their stable carbon site-specific isotope analysis". In: *Organic Geochemistry* 135, pp. 32–37. DOI: 10.1016/j.orggeochem.2019.06.002.

Chimiak, L. et al. (2021). "Carbon isotope evidence for the substrates and mechanisms of prebiotic synthesis in the early solar system". In: *Geochimica et Cosmochimica Acta* 292, pp. 188–202. DOI: 10.1016/j.gca.2020.09.026.

Clog, M. et al. (2018). "A reconnaissance study of ^{13}C–^{13}C clumping in ethane from natural gas". In: *Geochim Cosmochim Acta* 223, pp. 229–244.

Eigenbrode, J.L. et al. (2018). "Organic matter preserved in 3-billion-year-old mudstones at Gale crater, Mars". In: *Science* 360, pp. 1096–1101. DOI: 10.1126/science.aas9185.

Eiler, J.M. et al. (2017). "Analysis of molecular isotopic structures at high precision and accuracy". In: *International Journal of Mass Spectrometry* 422, pp. 126–142.

Freissinet, C. et al. (2015). "Organic molecules in the Sheepbed Mudstone, Gale Crater, Mars". In: *Journal of Geophysical Research: Planets*, pp. 495–514. DOI: 10.1002/2014JE004737.Received.

Gilmour, I. and C.T. Pillinger (1994). "Isotopic compositions of individual polycyclic aromatic hydrocarbons from the Murchison meteorite". In: *Mon. Not. R. Astron. Soc* 269, pp. 235–250.

Gordon, E.F. and D.C. Muddiman (2001). "Impact of ion cloud densities on the measurement of relative ion abundances in Fourier transform ion cyclotron resonance mass spectrometry: Experimental observations of coulombically induced cyclotron radius perturbations and ion cloud dephasing rates". In: *Journal of Mass Spectrometry* 36, pp. 195–203. DOI: 10.1002/jms.121.

Hayes, J.M. et al. (1990). "Compound-specific isotopic analyses: A novel tool for reconstruction of ancient biogeochemical processes". In: *Organic Geochemistry* 16, pp. 1115–1128.

Hoegg, E.D., C.J. Barinaga, et al. (2016). "Preliminary Figures of Merit for Isotope Ratio Measurements: The Liquid Sampling-Atmospheric Pressure Glow Discharge". In: *Journal of the American Society of Mass Spectrometry* 27, pp. 1393–1403. DOI: 10.1021/jasms.8b05326.

Hoegg, E.D., B.T. Manard, et al. (2019). "Initial Benchmarking of the Liquid Sampling-Atmospheric Pressure Glow Discharge-Orbitrap System Against Traditional Atomic Mass Spectrometry Techniques for Nuclear Applications". In: *Journal of the American Society for Mass Spectrometry* 30, pp. 278–288. DOI: 10.1007/s13361-018-2071-2.

Hoegg, E.D., R.K. Marcus, et al. (2018). "Concomitant ion effects on isotope ratio measurements with liquid sampling-atmospheric pressure glow discharge ion source Orbitrap mass spectrometry". In: *Journal of Analytical Atomic Spectrometry* 33, pp. 251–259. DOI: 10.1039/c7ja00308k.

Hofmann, A.E. et al. (2020). "Using Orbitrap mass spectrometry to assess the isotopic compositions of individual compounds in mixtures". In: *Int J Mass Spectrom* 457, p. 116410.

Huang, Y. et al. (2015). "Hydrogen and carbon isotopic ratios of polycyclic aromatic compounds in two CM2 carbonaceous chondrites and implications for prebiotic organic synthesis". In: *Earth Planet Sci Lett* 426, pp. 101–108.

Jang, C., L. Chen, and J.D. Rabinowitz (2018). "Metabolomics and Isotope Tracing". In: *Cell* 173, pp. 822–837. DOI: 10.1016/j.cell.2018.03.055.

Mahaffy, P.R. et al. (2012). "The sample analysis at Mars investigation and instrument suite". In: *Space Science Reviews* 170, pp. 401–478. DOI: 10.1007/s11214-012-9879-z.

Makarov, A. (2000). "Electrostatic Axially Harmonic Orbital Trapping: A High-Performance Technique of Mass Analysis". In: *Analytical Chemistry* 72, pp. 1156–1162.

Matthews, D.E., J.M. Hayes, and D.E. Matthews (1978). "Isotope-Ratio-Monitoring Gas Chromatography-Mass Spectrometry". In: *Analytical Chemistry* 50, pp. 1465–1473. DOI: 10.1021/ac50033a022.

Matucha, M. et al. (1991). "Isotope effect in gas-liquid chromatography of labelled compounds". In: *Journal of Chromatography A* 588, pp. 251–258. DOI: 10.1016/0021-9673(91)85030-J.

Mueller, E.P. et al. (2021). "Simultaneous, High-Precision Measurements of $\delta^2 H$ and $\delta^{1}3C$ in Nanomole Quantities of Acetate Using Electrospray Ionization-Quadrupole-Orbitrap Mass Spectrometry". In: *Analytical Chemistry.* DOI: 10.1021/acs.analchem.1c04141.

Naraoka, H., A. Shimoyama, and K. Harada (2000). "Isotopic evidence from an Antarctic carbonaceous chondrite for two reaction pathways of extraterrestrial PAH formation". In: *Earth Planet Sci Lett* 184, pp. 1–7.

Naraoka, H., Y. Yamashita, et al. (2017). "Molecular Evolution of N-Containing Cyclic Compounds in the Parent Body of the Murchison Meteorite". In: *ACS Earth and Space Chemistry* 1, pp. 540–550. DOI: 10.1021/acsearthspacechem.7b00058.

Neubauer, C., A. Cremiere, et al. (2020). "Stable Isotope Analysis of Intact Oxyanions Using Electrospray Quadrupole-Orbitrap Mass Spectrometry". In: *Analytical Chemistry.* DOI: 10.1021/acs.analchem.9b04486.

Neubauer, C., M.J. Sweredoski, et al. (2018). "Scanning the isotopic structure of molecules by tandem mass spectrometry". In: *International Journal of Mass Spectrometry* 434, pp. 276–286.

Niemann, H.B. et al. (2010). "Composition of Titan's lower atmosphere and simple surface volatiles as measured by the Cassini-Huygens probe gas chromatograph mass spectrometer experiment". In: *Journal of Geophysical Research E: Planets* 115, pp. 1–22. DOI: 10.1029/2010JE003659.

Ricci, M.P. et al. (1994). "Acquisition and processing of data for isotope-ratio-monitoring mass spectrometry". In: *Organic Geochemistry* 21, pp. 561–571. DOI: 10.1016/0146-6380(94)90002-7.

Sessions, Alex L. (2006). "Isotope-ratio detection for gas chromatography". In: *Journal of Separation Science* 29 (12), pp. 1946–1961. DOI: 10.1002/jssc.200600002.

Silverman, S.N. et al. (2021). "Practical considerations for amino acid isotope analysis". In: *Organic Geochemistry* 164, p. 104345. DOI: 10.1016/j.orggeochem.2021.104345.

Su, X., W. Lu, and J.D. Rabinowitz (2017). "Metabolite spectral accuracy on Orbitraps". In: *Analytical Chemistry* 89, pp. 5940–5948. DOI: 10.1021/acs.analchem.7b00396.

Thevis, M., O. Krug, and W. Schänzer (2006). "Mass spectrometric characterization of efaproxiral (RSR13) and its implementation into doping controls using liquid chromatography-atmospheric pressure ionization-tandem mass spectrometry". In: *Journal of Mass Spectrometry* 41, pp. 332–338. DOI: 10.1002/jms.993.

Uechi, G.T. and R.C. Dunbar (1992). "Space charge effects on relative peak heights in Fourier transform-ion cyclotron resonance spectra". In: *Journal of the American Society of Mass Spectrometry* 3, pp. 734–741.

vanHook, W.A. (1969). "Isotope Separation By Gas Chromatography". In: *Isotope Effects in Chemical Processes*. American Chemical Society.

Wilkes, E.B. et al. (2022). "Position specific carbon isotope analysis of serine by gas chromatography/Orbitrap mass spectrometry, and an application to plant metabolism". In: *Rapid Communications in Mass Spectrometry* 36. DOI: 10.1002/rcm.9347.

Williams, T.J. et al. (2021). "Roles of collisional dissociation modalities on spectral composition and isotope ratio measurement performance of the liquid sampling atmospheric pressure glow discharge / orbitrap mass spectrometer coupling". In: *International Journal of Mass Spectrometry* 464, p. 116572. DOI: 10.1016/j.ijms.2021.116572.

Zeichner, S.S. (2021). szeichner/DirectElution: DirectElutionv1.0.0. DOI: 10.22002/D1.2182.

Zeichner, S.S. et al. (2021). "Early plant organics increased global terrestrial mud deposition through enhanced flocculation". In: *Science* 371.6528, pp. 526–529. DOI: 10.1126/science.abd0379.

Para-xylene standard preparation

A 1% (v/v) stock solution of p-xylene was prepared by combining 10 µL of p-xylene (AlfaAesar LOT# Q28BO33, 99% purity) with 1 mL of GC-grade hexane (Honeywell Burdick and Jackson, 95% purity LOT# DY044-US). For analyses, 15 µL of this 1% solution was diluted in 1 mL of hexane; 1 mL of this dilute standard solution was pipetted into a GC vial for analysis, where 1 μL injections were used for replicate p-xylene measurements ($\sim$1.2 nmol per injection).

Serine standards: preparation, characterization, and derivatization

A serine working standard was prepared by homogenizing commercially available L-serine (Lot No. BCBS0964V, $\geq$ 99.5% purity, BioUltra, Sigma Aldrich; 'SERC0') with L-serine labeled with 99 atom % ^{13}C at the carboxyl (C-1) atom position (Sigma Aldrich) to achieve the standard material "SERC1." Aliquots of SERC0 and SERC1 were dissolved in millipore water, dripped into liquid nitrogen, freeze-dried, and powdered with a stir rod to ensure homogeneous composition.

GC derivatives of SERC0 and SERC1 were prepared by methylating the carboxyl group and trifluoroacetylating the hydroxyl and amine groups, producing N,O-bis(trifluoroacetyl) serine methyl ester (monoisotopic mass 311.0228 Da). The derivatization protocols were adapted from Corr et al., 2007 and performed at Caltech. To prepare each standard, 1 mg was weighed into a 2 mL GC vial and dissolved in 100μL of anhydrous methanol (Lot# SHBK9757). Standard vials were placed on ice, where 25μL of acetyl chloride (Lot# BCBW8067) was added dropwise. Vials were heated at 70°C for 1 hour, then gently dried under N$_2$.

Dichloromethane (DCM; 2×100μL) was added to the vials, then carefully evaporated under N$_2$. Samples were dissolved in 250μL of ethyl acetate, and 50μL of trifluoroacetic anhydride (TFAA; Lot# SHBK5942) added to the vials and heated at 110°C for 10 minutes, after which they were briefly dried under N$_2$. DCM (100μL) was added to the vials to help in eliminating residual derivatizing agent, then carefully evaporated under N$_2$. Derivatized standards were suspended in 1mL of hexane, then serially diluted to reach appropriate concentrations for introduction into the GC Orbitrap (i.e., 10s-100s pmols per 1μL injection).

The molecular-average isotopic compositions of SERC0 and SERC1 were measured by EA-IRMS at Caltech, standardized against international reference materials

(USGS-44, -45 glycines; USGS-73, -74 valines; caffeine; NIST-8542 sucrose). The mean $\delta^{13}C$ values ($\pm$ 1 σ_{SE}) were -32.22 $\pm$ 0.02‰ (n = 9, SERC0) and -20.35 $\pm$ 0.02‰ (n = 6, SERC1). The position-specific isotopic compositions of SERC1 at C-1, C-2, and C-3 positions, expressed relative to SERC0, were calculated using principles of mass balance by assuming that the measured difference in $\delta^{13}C$ between SERC0 and SERC1 was due entirely to ^{13}C label added at the C-1 atomic position (i.e., $\delta^{13}C_{C1,SERC0}$ of SERC1 = 36.8‰) and that C-2 and C-3 did not differ between SERC1 and SERC0 (i.e., $\delta^{13}C_{C2,SERC0}$ of SERC1 = $\delta^{13}C_{C3,SERC0}$ of SERC1 = 0‰).

PAH standards: preparation and characterization

Stock solutions of certified-reference-material grade pyrene (99.2%, Sigma-Aldrich TraceCERT®, Lot # BCBR5529V) and fluoranthene (99.5%, Sigma-Aldrich TraceCERT®, Lot # BCBL1694V) were made by dissolving each component in separate 1:1 (v/v) mixtures of hexane (Honeywell Burdick and Jackson, 95% purity) and acetone (EMD Omni-Solv, 99.7% purity). Four standard solutions of this type, corresponding to concentrations of approximately 100, 10, 2.5, and 1 ppm were created for each PAH. From these single-PAH solutions, four two-component mixtures were created by combining equal volumes of the same-concentration single-component standards (e.g., 1 ppm pyrene and 1 ppm fluoranthene solutions, mixed 1:1 v/v). One milliliter of each mixed standard was pipetted into a GC vial, which was capped and sealed with parafilm in preparation for analysis.

Molecular-average carbon isotope measurements of PAH standards were performed via combustion elemental analyzer-isotope ratio mass spectrometer (EA-IRMS) at Caltech on a EA IsoLinkTM combustion elemental analyzer system coupled to a Delta V Plus isotope ratio mass spectrometer (both from Thermo Scientific, Bremen, Germany) via a ConFlo IV Universal interface (Thermo Scientific, Bremen, Germany). The mean $\delta^{13}C_{VPBD}$ values ($\pm$ 1 σ_{SE}) were -25.68 $\pm$ 0.66‰ (n = 3, pyrene) and -25.69 $\pm$ 0.45‰ (n = 3, fluoranthene).

Meteorite sampling and solution preparation

A subsample of Murchison, a CM2 carbonaceous chondrite meteorite, procured from a larger fragment supplied by the Field Museum was collected in a positive-pressure, HEPA-filtered laminar flow hood as follows: A $\sim$20 mm^2 area of the outer surface was first scraped away using a clean steel file. The file was then cleaned via sonication in high-purity methanol and used to further abrade the same surface.

This material was set aside. The file was re-cleaned using the same method with fresh methanol and then used to remove ~35 mg from the cleared-off meteorite surface for analysis. The powdered sample was collected and stored in a clean glass vial. Similar meteorite subsampling and collection measures are common preparation for organic extraction of PAHs (Aponte, Dworkin, and Elsila, 2015; Gilmour and Pillinger, 1994; Huang et al., 2015; Naraoka, Shimoyama, and Harada, 2000; Naraoka, Yamashita, et al., 2017).

Approximately 2.5 mg of the powdered meteorite were transferred into a glass volatile organic analysis (VOA) vial, to which 5 mL of a 1:1 v/v solvent mixture of high-purity hexane and acetone were added. The vial was loosely capped, placed in a 60°C water bath, and sonicated for 30 minutes. To remove particulates, the solvent phase was pipetted into a pre-washed 4"-long column filled with 150 Å silica beads over packed glass wool and collected in a clean beaker. An additional 3 mL of clean solvent mixture were run through the same column and collected. The beaker was loosely covered with aluminum foil and left in the fume hood until all but ~1 mL of solution had evaporated. This solution was transferred to a clean GC vial and dried down under flowing N_2 at 25°C until only ~0.1 mL remained. The vial was capped and sealed with parafilm in preparation for analysis.

Additional details of direct elution data analysis

We compared the data processing method outlined in "Direct elution modeling" to isotope ratio calculations calculated on NL scores alone (i.e., without conversion to ion counts before summing and computing ratios), as well as on integrated peak areas calculated by weighting ion counts for each scan by absolute intensity, and found that these other methods of computing ratios consistently generated less precise and less accurate results than the method used and applied in this study. For the intensity-weighted isotope ratios in particular, we found that they were neither accurate nor precise, because the weighting combined with the chromatographic shift of the isotope ratios over time preferentially favored the ratios in the middle of the peak elution, which are not necessarily representative of the absolute isotope ratio (Fig. S2.1).

We explored two different ways to calibrate the isotope ratios of samples based on standards measured sequentially during a single analytical session (generally the same day or small number of continuous days), in order to account for potential drift in instrumental conditions. All results presented in this study are based on the following approach: We first perform a linear regression on the measured isotope ratios of standards as a function of analytical sequence, and then interpolate or extrapolate that relationship to compute the expected isotope ratio of the standard at the run number when the sample was measured. We compared this approach to sample-standard comparison using the average of only the two time-adjacent standards and found that our preferred approach best accounted for any instrumental drift over time and produced the best relative standard errors of replicate analyses of samples and standards.

Chromatographic peak model details

We model the time-evolution of the intensity of the chromatographic peaks for the un-substituted and isotope-substituted forms of the eluting compounds as Gaussian functions:

$$y = a * exp\frac{-(t - t_o)^2}{2\sigma^2_{skewed}} + c \tag{S2.1}$$

where y is the intensity at time t, a is the maximum intensity of the peak, t is an array representing the time frame the peak is eluting (seconds), t_o is the elution time of the peak maximum, σ_{skewed} is the width (standard deviation) of each Gaussian peak, and c is the background (intensity at some time far from t_o). Within the variable σ_{skewed}, we incorporate tailing into the peaks via the linear function:

$$\sigma_{skewed} = \sigma + s * t \tag{S2.2}$$

where σ is the standard deviation of the peak, s is the peak skew, and t is the time array.

In practice, the fluence of ions observed by the Orbitrap is not continuous; the C-trap fills with ions until it reaches a set injection time (IT; defined by the previous scan) where it expects to reach a critical threshold of ions, after which it introduces the accumulated packet of ions into the Orbitrap for analysis. The time period that a packet of ions is observed by the Orbitrap mass analyzer is constant from scan to scan, a function of the resolution defined for the measurement (~300ms for 120k), and often longer than the time it takes for the C-trap to reach capacity (i.e., AGC target). The IT is recalculated based on how many ions were observed in the previous scan; if the computed IT time is shorter than the length of the Orbitrap scan observation time, the C-trap can start re-filling prior to the end of the analysis of the previous ion packet by the Orbitrap, so that it can introduce a new packet of ions into the Orbitrap when the previous scan concludes. The timing of when the C-trap starts filling again is calculated based on the ion flux in the last scan, and how long it takes that given ion fluence to reach the AGC target (i.e., longer at the tails of a peak and shorter at the peak maximum). The AGC target is approximately equal to the total ion current (TIC) times the injection time (IT) for that scan (i.e., TIC×IT).

In our model, we model this C-trap ion accumulation by counting the number of ions accumulated since a scan begins (TIC) and multiplying the TIC by the width of each time step (0.1 millisecond in our model). If it is not the first time step in the scan, this value gets added to the sum calculated in previous scans, which represents the current AGC threshold. In our model, we simplify $TIC = a_{substituted} + a_{unsubstituted}$, although in reality there are other background fragment ions contributing to the TIC. Once the C-trap has observed the ion fluence for the length of the IT time, the ions are injected into the Orbitrap. The difference in the number of ions seen by the Orbitrap mass analyzer compared to the number of sample ions introduced into the system defines the sensitivity of the experiment (Eiler et al., 2017).

Each of variables in Eqns. S2.1 and S2.2, as well as the discontinuous time-based observation of ion fluence by the Orbitrap can have a potential effect on measured isotope ratios. We choose standard parameters for chromatographic peak intensities (NL scores; 1×10^9), relative peak abundance, max IT time, resolution (120k) and AGC target (2×10^5). The complete code for this model can be found on the Caltech Data repository (Zeichner, 2021), and could serve as a useful tool to design future direct elution experiments.

We iterate through parameter combinations within our numerical model as follows to estimate the potential sources of error in our measurement and data analysis methods for a single analyte isotope ratio measurement: the time offset of the chromatographic peaks (the offset between to for elution of the substituted and unsubstituted species (0-150 ms)), the peak width (σ = 1-6 s), the skew of the peak (s = 0.05-0.2), and fraction of maximum peak intensity at which we define the baseline to begin and end peak integration (0-0.3). We used the same peak width and skew parameters for both the substituted and unsubstituted fragment ions of a given model run, based on an assumption that the chromatographic column would produce the same peak shapes for both. For each modeled peak, we also added a background, c, that was 3 orders of magnitude less than the NL score of the ion species of interest ($a \times 10^{-3}$). This Monte Carlo approach was used to gain an understanding of which parameters could have the largest effect on single analyte isotope ratio measurements.

In addition, we applied our model to evaluate the potential effects of differences in isotope abundance between a sample and standard ($\frac{a_{substituted-sample}}{a_{substituted-standard}} = 0.001, 0.01,$ 0.1, 1), as well as differences in their measured concentrations (i.e., moles of analyte) on the accuracy and precision for a sample-standard comparison (i.e., δ value). For these model runs, we also choose standard parameters for chromatographic peak intensities: NL scores; 1×10^{9}, with $\sigma = 3s$, skew = 0.1, peak offset = 50 ms, backgrounds that are 3 orders of magnitude less than the NL score of the ion species of interest, max IT time=3000 ms, resolution= 120k and AGC target = 2×10^{5}. Results of these model runs are presented in Fig. 2.6 and discussed further in the main text.

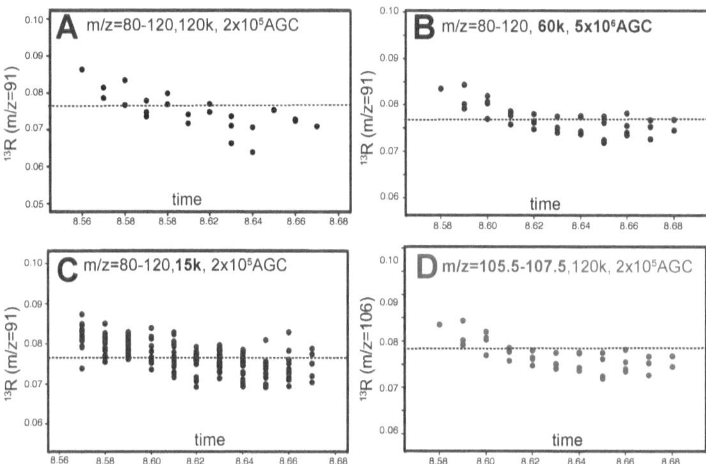

Fig. S2.1: Para-xylene acquisitions and replicate standard errors.

Single acquisitions for four of the para-xylene experiments depicted in Fig. 2.3, demonstrating the $^{13/12}R$ for 91 (A-C) and 106 (D) mass spectral peaks. Dotted lines demonstrate average $^{13/12}R$ across replicate acquisitions compared to $^{13/12}R$ over time within each acquisition; grey bar represents RSE across replicate acquisitions.

Term	Abbreviation	Definition
Acquisition	n/a	Single sample injection into the Orbitrap
Automatic gain control	AGC	System control responsible for admitting a constant number of ions into the mass analyzer with each acquisition, which should be ~TIC×IT. Operating under "AGC control" means that TIC×IT is stable (variation <0.2) for all scans within an acquisition.
AGC target	n/a	The defined number of ions to admit into the Orbitrap within each scan (e.g., 2×10^5)
Advanced Quadrupole System	AQS	Portion of the mass spectrometer responsible for mass filtering the ions entering the Orbitrap
C trap	n/a	Portion of the mass spectrometer where the ions are accumulated based on the set AGC target and stored prior to introduction into the Orbitrap mass analyzer
Injection time	IT	Length of time that the C trap introduces ions into the Orbitrap
NL score	n/a	Orbitrap reported signal intensity, which can be converted into ion counts for isotope ratio analysis
Scan	n/a	One observational cycle in the Orbitrap, which produces a transient that undergoes a fast Fourier transform into mass spectrometric data. The length of the scan is set by resolution.
Total ion current	TIC	The total number of ions observed by the Orbitrap in a given scan
Transient	n/a	Raw data product from the Orbitrap prior to fast Fourier transform

Table S2.1: Glossary of Terms.

Terms used with regards to Orbitrap isotope ratio analysis, and their associated abbreviations.

Table S2.2: Full list of direct injection experiments. See page 54 for caption.

Comp.	Source (LOT#)	Experiment goal	Isotope ratio target (mass)	Mass window	GC oven program	Time frame of elution	AGC target	Res.	Isotope ratio (R)	σ_{RSE} (σ_{SNL}/R)
Para-xylene	AlfaAesar (LOT# Q28BO33)	Characterize precision of large mass window, high resolution, moderate AGC	$^{13/12}$C (m/z=91, 106)	80-120	50 to 100° by 5°min^{-1}, 5 min 100°hold	8.54-8.86	2e5	120	$^{13/12}$C$_{91}$= 0.00767; $^{13/12}$C$_{106}$ = 0.0867	$\sigma_{RSE,91}$ = 3.1‰ (σ_{SNL}/R =0.0054); $\sigma_{RSE,106}$ = 3.4‰ (σ_{SNL}/R =0.0074)
Para-xylene	AlfaAesar (LOT# Q28BO33)	Characterize precision of large mass window, low resolution, moderate AGC	$^{13/12}$C (m/z=91, 106)	80-120	"	8.54-8.86	2e5	15	$^{13/12}$C$_{91}$ = 0.0769; $^{13/12}$C$_{106}$ = 0.0885	$\sigma_{RSE,91}$ = 1.3‰ ($\sigma_{SNL,91}$/R = 0.0021); $\sigma_{RSE,106}$ = 2.1‰ ($\sigma_{SNL,106}$/R =0.0028)

Continued on next page

Table S2.2: Full list of direct injection experiments. See page 54 for caption. (Continued)

Para-xylene	AlfaAesar (LOT# Q28BO33)	Characterize precision of large mass window, moderate resolution, high AGC	$^{13/12}$C (m/z=91, 106)	80-120	"	8.54-8.86	5e6	60	$^{13/12}C_{91}$ = 0.0769; $^{13/12}C_{106}$ = 0.0895	$\sigma_{RSE,91}$ = 2.2‰ ($\sigma_{SNL,91}/R$ = 0.0022); $\sigma_{RSE,106}$= 2.2‰ ($\sigma_{SNL,106}/R$ =0.0018)
Para-xylene	AlfaAesar (LOT# Q28BO33)	Characterize precision of small mass window for ^{13}C and D	$^{13/12}$C (m/z=106), $^{2/1}$D (m/z=106)*	105.5-107.5	"	8.54-8.86	2e5	120	$^{13/12}C_1$06 = 0.0786; $^{2/1}D_{106}$ = 0.012	σ_{RSE-C} = 5.4‰ (σ_{SNL-C}/R = 0.0047); σ_{RSE-D} = 107‰ (σ_{SNL-D}/R =0.041)
Para-xylene	AlfaAesar (LOT# Q28BO33)	Characterize precision limits of M+1 fragment	$^{2/13}$C (m/z=106)*107.5	106.5-107.5	"	8.54-8.86	5e4	120	$^{2/13}C_{106}$ = 0.0117	$\sigma_{RSE-D/C}$ = 17.9‰ ($\sigma_{SNL-D/C}/R$ = 0.017.8)

Continued on next page

Table S2.2: Full list of direct injection experiments. See page 54 for caption. (Continued)

Fluor.		$^{13/12}C$ and $2x^{13/12}C$ of molecular ion	$^{13/12}C$ (m/z=202), $^{13/12}C^{13/12}C$ (m/z=202)	198-206	1min 60°hold, 60 to 225° by 6° min^{-1}, 11.5 min hold	32.65-33.35	2e5	180	$^{13/12}C_{202}$ = 0.1690; $^{13/12}C$ $^{13/12}C_{202}$ = 0.0119	$\sigma_{RSE-13/12}$ = 1.9‰ ($\sigma_{SNL-13/12}$ /R = 0.0041); $\sigma_{RSE-2x13/12}$ = 8.7‰ ($\sigma_{SNL-2x13/12}$/R2=0.014)
Pyrene		$^{13/12}C$ and $2x^{13/12}C$ of molecular ion	$^{13/12}C$ (m/z=202), $^{13/12}C^{13/12}C$ (m/z=202)	198-206	"	34.15-34.80	2e5	180	$^{13/12}C_{202}$ = 0.1664; $^{13/12}C$ $^{13/12}C_{202}$ = 0.0119	$\sigma_{RSE-13/12}$ = 1.7‰ ($\sigma_{SNL-13/12}$ /R = 0.0041); $\sigma_{RSE-2x13/12}$ = 9.5‰ ($\sigma_{SNL-2x13/12}$/R =0.014)
Serine (SERC0)	Lot No. BCBS0964V, 99.5% purity, BioUltra, Sigma Aldrich	$^{13/12}C$ of 110, 138 and 165 fragments	$^{13/12}C$ (m/z=110, 138, 165)	105-180	50 to 85° by 15° min^{-1}, 15 min 85°hold	9.0-10.0	2e5	120	$^{13/12}C_{110}$ = 0.0308, $^{13/12}C_{138}$ = 0.0422, $^{13/12}C_{165}$ = 0.0556	$\sigma_{RSE-110}$ = 3.7‰ ($\sigma_{SNL-110}$/R = 0.0077); $\sigma_{RSE-138}$ = 7.1‰ ($\sigma_{SNL-138}$ /R =0.0047); $\sigma_{RSE-165}$ = 1.0‰ ($\sigma_{SNL-165}$/R =0.0115)

Continued on next page

Table S2.2: Full list of direct injection experiments. See page 54 for caption. (Continued)

Serine (SERC1)	Lot No. BCBS0964V + L-serine C1 label (99% purity)	$^{13/12}$C of 110, 138 and 165 fragments	$^{13/12}$C (m/z=110, 138, 165)	"	"	9.0-10.0	"	"	$^{13/12}$C$_{110}$ = 0.0311, $^{13/12}$C$_{138}$ = 0.0418, $^{13/12}$C$_{165}$ = 0.0559	$\sigma_{RSE-110}$ = 3.1‰ ($\sigma_{SNL-110}$/R = 0.0081); $\sigma_{RSE-138}$ = 6.9‰ ($\sigma_{SNL-138}$/R =0.0050); $\sigma_{RSE-165}$ = 14.7 ‰ ($\sigma_{SNL-165}$/R =0.0125)
Serine (SERC0)	Lot No. BCBS0964V, 99.5% purity, BioUltra, Sigma Aldrich	$^{13/12}$C of 165 fragment	$^{13/12}$C (m/z=165)	162-169	50 to 85°by 15min, 15 min 85° hold	9.0-10.0	2e5	120	$^{13/12}$C$_{165}$ = 0.0537	$\sigma_{RSE-165}$ = 2.0‰ ($\sigma_{SNL-165}$/R =0.0042)
Serine (SERC1)	Lot No. BCBS0964V + L-serine C1 label (99% purity)	$^{13/12}$C of 165 fragment	$^{13/12}$C (m/z=165)	"	"	89.0-10.0	"	"	$^{13/12}$C$_{165}$ = 0.0557	$\sigma_{RSE-165}$ = 2.0‰ ($\sigma_{SNL-165}$/R =0.0040)

Table S2.2: Full list of direct injection experiments.

Standards with source information were injected with a range of GC oven programs and Orbitrap conditions, optimized to the substrate of interest. We report calculated isotope ratios (R) with reported relative standard errors (σ_{RSE}) compared to their shot noise limits (σ_{SNL}). Experiments with asterisks next to them were not baseline corrected. Fluor = fluoranthene.

	sample	$\delta^{13}C_{VPDB}$ (‰)	σ_{RSE} of $\delta^{13}C_{VPDB}$ (‰)	$^{2x13/12}R$	σ_{SE} ($^{2x13/12}R$)	conversion of measured $^{2x13/12}R$ to corrected ratio ($^{2x13/12}R*$)
This study	Murchison pyrene	-11.81	1.8	0.01210072	0.00011545	0.01269073
	Murchison fluoranthene	-3.92	1.9	0.01269114	0.0001099	0.01328115
	Standard pyrene	-25.7	0.66	0.01188313	–	0.01247314
	Standard fluoranthene	-25.7	0.45	0.01186934	–	0.01245935
Naraoka, Shimoyama, and Harada, 2000	Murchison pyrene	-13.1	1.3	–	–	–
	Murchison fluoranthene	-5.9	1.1	–	–	–
	Standard pyrene	-23.6	1.3	–	–	–
	Standard fluoranthene	-24.7	1.0	–	–	–

Table S2.3: Full list of Murchison PAH singly- and doubly- substituted carbon isotope measurements from this study, and Naraoka, Shimoyama, and Harada, 2000.

POLYCYCLIC AROMATIC HYDROCARBONS IN SAMPLES OF RYUGU FORMED IN THE INTERSTELLAR MEDIUM

S.S. Zeichner, J.C. Aponte, et al. (2023). "Polycyclic aromatic hydrocarbons in samples of Ryugu formed in the interstellar medium". In: *Science*. DOI: `10.1126/science.adg6304`.

Sarah Zeichner[1], José C. Aponte[2], Surjyendu Bhattacharjee[1], Guannan Dong[1], Amy E. Hofmann[1,3], Jason P. Dworkin[2], Daniel P. Glavin[2], Jamie E. Elsila[2], Heather V. Graham[2], Hiroshi Naraoka[4], Yoshinori Takano[5], Shogo Tachibana[6,7], Allison T. Karp[8,9,10], Kliti Grice[11], Alex I. Holman[11], Katherine H. Freeman[8], Hisayoshi Yurimoto[12], Tomoki Nakamura[13], Takaaki Noguchi[14], Ryuji Okazaki[4], Hikaru Yabuta[15‡], Kanako Sakamoto[7], Toru Yada[7], Masahiro Nishimura[7], Aiko Nakato[7], Akiko Miyazaki[7], Kasumi Yogata[7], Masanao Abe[7,16] Tatsuaki Okada[6,17], Tomohiro Usui[6,7], Makoto Yoshikawa[7,16], Takanao Saiki[7], Satoshi Tanaka[7,16], Fuyuto Terui[18], Satoru Nakazawa[7], Sei-ichiro Watanabe[19], Yuichi Tsuda[7], Kenji Hamase[20], Kazuhiko Fukushima[21], Dan Aoki[21], Minako Hashiguchi[19], Hajime Mita[22], Yoshito Chikaraishi[23], Naohiko Ohkouchi[5], Nanako O. Ogawa[5], Saburo Sakai[5], Eric T. Parker[2], Hannah L. McLain[2], Francois-Regis Orthous-Daunay[24], Véronique Vuitton[24], Cédric Wolters[24], Philippe Schmitt-Kopplin[25,26,27], Norbert Hertkorn[25,28], Roland Thissen[29], Alexander Ruf[30,31,32], Junko Isa[33,34], Yasuhiro Oba[23], Toshiki Koga[5], Toshihiro Yoshimura[5], Daisuke Araoka[35], Haruna Sugahara[7], Aogu Furusho[36†], Yoshihiro Furukawa[13], Junken Aoki[37], Kuniyuki Kano[37], Shin-ichiro M. Nomura[38], Kazunori Sasaki[39,40], Hajime Sato[39], Takaaki Yoshikawa[41], Satoru Tanaka[42], Mayu Morita[42], Morihiko Onose[42], Fumie Kabashima[43], Kosuke Fujishima[31,44], Tomoya Yamazaki[23], Yuki Kimura[23], John M. Eiler[1]

Affiliations: [1]Geological and Planetary Science Division, California Institute of Technology, Pasadena, CA 91125, USA, [2]Solar System Exploration Division, NASA Goddard Space Flight Center, Greenbelt, MD 20771, USA, [3]Jet Propulsion Laboratory, California Institute of Technology, Pasadena, CA 91109, USA, [4]Department of Earth and Planetary Sciences, Kyushu University, Fukuoka 819-0395, Japan, [5]Japanese Agency for Marine-Earth Science and Technology, Yokosuka, Kanagawa, 237-0061, Japan, [6]Department of Earth and Planetary Science, University of Tokyo, Tokyo 113-0033, Japan, [7]Institute of Space and Astronautical Science, Japan Aerospace Exploration Agency, Sagamihara 252-5210, Japan, [8]Department of Geosciences, The Pennsylvania State University, University Park, PA 16802, USA, [9]Ecology and Evolutionary Biology Department, Yale University, New Haven, CT, USA, [10]Department of Environmental, Earth, and Planetary Sciences, Brown University, Providence, RI 02912, USA, [11]Western Australia Organic & Isotope Geochemistry Centre, The Institute for Geoscience Research, School of Earth and Planetary Sciences, Curtin University, Perth, Western Australia 6102, Australia, [12]Department of Earth and Planetary Sciences, Hokkaido University, Sapporo 060-0810, Japan, [13]Department of Earth Science, Tohoku University, Sendai 980-8578, Japan, [14]Division of Earth and Planetary Sciences, Kyoto University, Kyoto 606-8502, Japan, [15]Department of Earth and Planetary Systems Science, Hiroshima University, Higashi-Hiroshima 739-8526, Japan, [16]School of Physical Sciences, The Graduate University for Advanced Studies, Hayama 240-0193, Japan, [17]Department of Chemistry, University of Tokyo, Tokyo 113-0033, Japan, [18]Department of Mechanical Engineering, Kanagawa Institute of Technology, Atsugi 243-0292, Japan, [19]Graduate School of Environmental Studies, Nagoya University, Nagoya 464-8601, Japan, [20]Graduate School of Pharmaceutical Sciences, Kyushu University, Fukuoka 812-8582, Japan, [21]Graduate School of Bioagricultural Sciences, Nagoya University, Nagoya 464-8601, Japan, [22]Department of Life, Environment and Material Science, Fukuoka Institute of Technology, Fukuoka 811-0295, Japan, [23]Institute of Low Temperature Science, Hokkaido University, Sapporo 060-0189, Japan, [24]Institut de Planétologie et d'Astrophysique de Grenoble, Université Grenoble Alpes, Centre National de la Recherche Scientifique, 38000 Grenoble, France, [25]Analytical BioGeoChemistry, Helmholtz Zentrum München, 85764 Neuherberg, Germany, [26]Technische Universität München, Analytische Lebensmittel Chemie, 85354 Freising, Germany, [27]Max Planck Institute for Extraterrestrial Physics, 85748 Garching bei München, Germany, [28]Department of Thematic Studies,

Environmental Sciences, Linköping University, 58183 Linköping, Sweden, [29]Institut de Chimie Physique, Université Paris-Saclay, Centre National de la Recherche Scientifique, Orsay 91405, France, [30]Laboratoire de Physique des Interactions Ioniques et Moléculaires, Université Aix-Marseille, Centre National de la Recherche Scientifique, 13397 Marseille, France, [31]Faculty of Physics, Ludwig-Maximilians-University, 80799 Munich, Germany, [32]Excellence Cluster Origins, 85748 Garching, Germany, [33]Earth-Life Science Institute, Tokyo Institute of Technology, Tokyo 1528550, Japan, [34]Planetary Exploration Research Center, Chiba Institute of Technology, Narashino 275-0016, Japan, [35]Geological Survey of Japan, National Institute of Advanced Industrial Science and Technology, Tsukuba, Ibaraki 305-8567, Japan, [36]Graduate School of Pharmaceutical Sciences, Kyushu University, Fukuoka 819-0395, Japan, [37]Graduate School of Pharmaceutical Sciences, University of Tokyo, Tokyo 113-0033, Japan, [38]Department of Robotics, Tohoku University, Sendai 980-8578, Japan, [39]Human Metabolome Technologies Inc., Kakuganji, Tsuruoka, Yamagata, 997-0052, Japan, [40]Institute for Advanced Biosciences (IAB), Keio University, Kakuganji, Tsuruoka, Yamagata, 997-0052, Japan, [41]Horiba Advanced Technologies, Co., Ltd. Kisshoin, Minami-ku, Kyoto 601-8510, Japan, [42]HOriba Technonology Services Co., Ltd. Kisshoin, Minami-ku, Kyoto 601-8510, Japan, [43]Laboratory Equipment Corporation Japan, Tokyo 105-0014, Japan, [44]Graduate School of Media and Governance, Keio University, Fujisawa, Kanagawa, 252-0882, Japan

‡Present address: International Institute for Sustainability with Knotted Chiral Meta Matter, Hiroshima University, Hiroshima 739-8526, Japan.

†Present address: School of Pharmaceutical Sciences, University of Shizuoka, Shizuoka 422-8526, Japan.

Polycyclic aromatic hydrocarbons (PAHs)—organic molecules consisting of multiple aromatic rings—are ubiquitous in the interstellar medium (ISM). Based on observations of mid-infrared emission bands in the ISM, PAHs are present in abundances $\sim 10^{-7}$ times that of hydrogen (Tielens, 2013). PAHs are estimated to contain 20% of the carbon atoms in the ISM of the Milky Way (Allamandola, Sandford, and Wopenka, 1987; Tielens, 2013) and other galaxies (Smith et al., 2007). PAHs have been proposed as building blocks of carbon-rich dust grains, which are abundant in the ISM (Draine, 2016), and of higher molecular weight insoluble organic material (IOM) that comprises most of the carbon within meteorites (Tielens, 2008). However, it is unknown which chemical processes produce these forms of reduced carbon, or where they occur (Fig. 3.1) (Kaiser and Hansen, 2021).

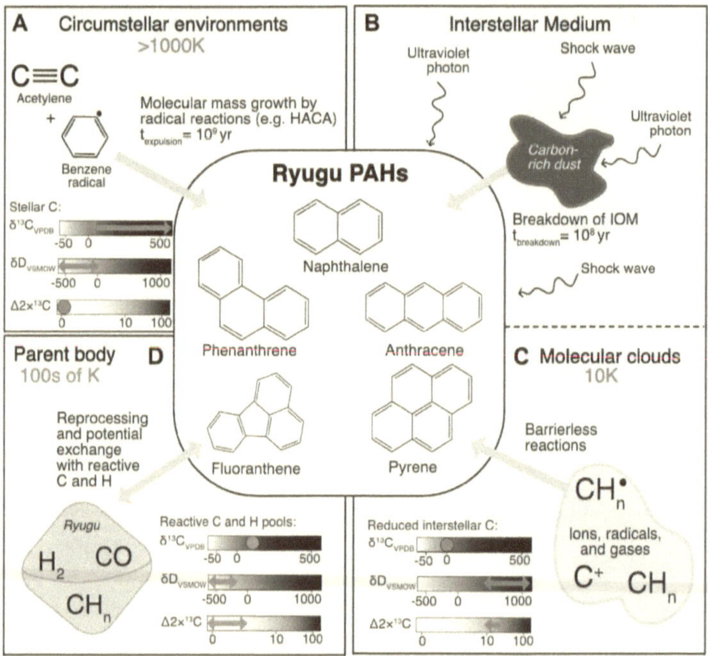

Figure 3.1: Potential pathways for extraterrestrial PAH formation.

Caption continued on next page.

Figure 3.1: Potential pathways for extraterrestrial PAH formation.

The central inset shows the molecular structures of the five PAHs we investigated. The surrounding panels schematically illustrate potential formation pathways for those PAHs. In panels A, C and D, grayscale color bars show $\delta^{13}C_{VPBD}$, δD_{VSMOW} and $\Delta2\times^{13}C$ values measured or predicted in extraterrestrial materials. White is isotopically depleted and black is isotopically enriched. Dots and arrows indicate values and ranges, respectively, of the source carbon and hydrogen. $\Delta2\times^{13}C$ values are estimated predictions (Supplemental Methods) based on model results (Fig. 3.2, see text). (A) PAH formation in hot ($\gtrsim$1000K, red) circumstellar environments by molecular mass growth reactions (Kaiser and Hansen, 2021; Micelotta, Jones, and Tielens, 2010a). $\delta^{13}C_{VPBD}$ values in AGB stars are expected to range from 0 to hundreds of ‰ depending on the stellar evolution (Abia et al., 2003; Busso, Gallino, and Wasserburg, 1999). δD_{VSMOW} values are expected to be low or zero due to the fusion of D in stars (Abia et al., 2003; Busso, Gallino, and Wasserburg, 1999). $t_{expulsion}$ is the times cale for PAH expulsion from stellar envelopes. (B) Shock waves and ultraviolet radiation form PAHs by breaking down carbon-rich dust, but can also destroy PAHs. $t_{breakdown}$ is the time scale for PAH breakdown in the ISM. (C) PAH formation in cold (10K, blue) interstellar environments through barrierless reactions (Doddipatla et al., 2020; Kaiser and Hansen, 2021). Reduced carbon in molecular clouds is depleted in ^{13}C contents compared to interstellar CO (Cordiner et al., 2019), while interstellar hydrogen is typically D-enriched (Sandford, 2002). (D) PAH formation or modification on a parent body at moderate temperatures (100s of K, orange). Isotopic exchange can occur with carbon reservoirs such as CO and DIC, and hydrogen reservoirs such as H_2. Murchison carbonate has $\delta^{13}C_{VPBD}$ values of +20 to +80‰ (Sephton, 2002), while the water in parent bodies of CC meteorites has been found to be D-depleted (Alexander, Bowden, Fogel, Howard, et al., 2012; McCollom et al., 2010).

Small aromatic organics, such as PAHs containing only a few rings, can form through reactions of free radicals in the gas phase, particularly the hydrogen-abstraction-carbon-addition (HACA) reaction mechanism which is expected to occur in hot ($\geq$1000K) circumstellar environments around carbon-rich asymptotic giant branch (AGB) stars and on Earth by combustion (Fig. 3.1A) (Kaiser and Hansen, 2021). Carbon-rich dust grains and IOM could potentially be formed via similar processes. However, the reaction rates of these high-temperature mechanisms are too slow to account for the amount of PAHs present within the ISM and there is no complete model of the synthesis of PAHs within the outflows of AGB stars (Kaiser and Hansen, 2021; Kaiser, Parker, and Mebel, 2015).

PAHs could also be formed through the breakdown of carbon-rich dust grains by shock waves, cosmic rays, or ultraviolet photolysis (Chiar et al., 2013). However, these same processes destroy PAHs (Fig. 3.1B). This destruction occurs on time scales (Micelotta, Jones, and Tielens, 2010a,b) that are shorter than expected for production of PAHs in circumstellar envelopes of AGB stars (Doddipatla et al., 2020; Frenklach and Feigelson, 1989).

A third location where PAHs could form is in cold ($\sim$10K) molecular clouds within the ISM through either ion-molecule reactions (Martinez et al., 2008), or rapid barrierless reactions involving radicals (Fig. 3.1C) (Kaiser, Parker, and Mebel, 2015). Laboratory experiments have characterized these chemical mechanisms, but it is difficult to directly observe specific PAH molecules within interstellar molecular clouds using spectroscopic methods. The only species that have been identified within molecular clouds are nitriles derived from PAHs: benzo-nitrile (McGuire, Burkhardt, et al., 2018) and cyanonaphthalenes (McGuire, Loomis, et al., 2021). It is therefore unlikely that circumstellar synthesis dominates the formation of extraterrestrial PAHs, but there is little evidence for interstellar formation either.

Later secondary processing reactions within a parent body—the asteroid or other Solar System object that meteorites originate from—could also synthesize PAHs or alter their composition. These reactions are often related to aqueous alteration, which is the modification of solid material by reactions with liquid water, which is known to have occurred on parent bodies. Potential secondary reactions include Fischer-Trospch-type (FTT) synthesis of alkanes from carbon monoxide (CO) (Zolotov and Shock, 1999) followed by aromatization; exchange with dissolved inorganic carbon (DIC) (Graham et al., 2022) or aqueous H_2; or the breakdown of larger macromolecular insoluble organic matter (IOM) into smaller organic molecules by catagenesis (Pehr et al., 2021), which is a process that thermally cracks large organic molecules (on Earth this forms oil and gas deposits).

Soluble PAHs have been studied in samples of carbonaceous chondrite (CC) meteorites, but due to the terrestrial exposure of meteorite samples, it is possible that endogenous PAHs in meteorites could be contaminated with PAHs formed on Earth. Samples of the near-Earth carbonaceous asteroid (162173) Ryugu were collected by the Hayabusa2 spacecraft under controlled conditions, so underwent fewer opportunities for terrestrial contamination. We studied the isotopic properties of PAHs in the CC meteorite Murchison and a sample of the asteroid Ryugu to investigate PAH formation processes.

Principles of isotope analysis

The rare-isotope distributions of organic molecules can be used to constrain the source, substrate, and chemical mechanisms responsible for their formation (Bigeleisen and Wolfsberg, 1957; Eiler, 2013). The molecular-average $^{13}C/^{12}C$ and D/H ratios of organic molecules are conventionally reported as delta values $\delta^{13}C_{VPBD}$ and δD_{VSMOW}, respectively, where δ notation is defined in Eqn. S3.7, and VPDB and VSMOW are standard reference scales (See Supplemental Materials and Methods). Measured isotope ratios of extraterrestrial organics are often outside the ranges of ratios measured in terrestrial organics. Some extraterrestrial organic molecules, such as amino acids (Naraoka et al., 2023), are enriched in ^{13}C and have been interpreted as products of precursor molecules that were themselves derived from a reservoir of CO in the ISM that was enriched in ^{13}C (Elsila et al., 2012; Glavin et al., 2018). In contrast, interstellar PAHs exhibit only small ^{13}C enrichments, consistent with formation in a range of extraterrestrial environments (Fig. 3.1). Heterogeneities in the carbon isotopic compositions of interstellar, circumstellar, and parent body carbon reservoirs prevent unique interpretation of such data (Fig. 3.1).

Multiple substitutions of heavy isotopes within molecules (hereafter referred to as clumping) reflect temperature-dependent chemical reactions and physical processes (Eiler, 2013), and so provide additional information on the molecule's source and formation/degradation history. For any given molecular-average $\delta^{13}C_{VPBD}$ value of PAHs, a smaller proportion of each molecule is doubly-^{13}C substituted (denoted $2\times^{13}C$). We report the ratio '$2\times^{13}C/^{12}C$' of clumped ^{13}C isotopologues to the unsubstituted isotopologues, and also the differences– $\Delta2\times^{13}C$– values–between the measured $2\times^{13}C/^{12}C$ ratios and the expected stochastic $2\times^{13}C/^{12}C$ ratios. Expected $2\times^{13}C/^{12}C$ ratios are calculated based on the random probability of a double-^{13}C-substitution given its molecular-average ^{13}C abundance (See Supplemental Materials and Methods). A formation process that leads to a higher (lower) number of heavy isotope substitutions than the stochastic expectation leads to a positive (negative) $\Delta2\times^{13}C$ value.

Variations in $2\times^{13}C$ clumping can be temperature-dependent, because heavy isotopic substitution lowers the molecular vibrational energy (Fig. S3.2) of the C-C bond and therefore stabilizes it. Isotopic clumping therefore is more energetically favored at lower temperatures. We used acetylene as a model compound to predict $2\times^{13}C$ clumping at different temperatures, because it is thought to be a precursor to the formation of interstellar aromatic reduced carbon (Lahuis and Dishoeck, 2000) including PAHs (Abplanalp and Kaiser, 2020; Mebel, Landera, and Kaiser, 2017). The equilibrium exchange reaction between single and double-^{13}C substitution in acetylene is:

$$2 \times^{13} C^{12}CH_2 \rightleftharpoons ^{13}C_2H_2 +^{12} C_2H2 \tag{3.1}$$

The products on the right-hand side of equation 3.1 are favored at all temperatures (Fig. 3.2; See Supplemental Materials and Methods). We therefore predict that if PAHs formed in the cold ISM, they will have $\Delta2\times^{13}$C values substantially different from zero (positive if formed by reversible reactions; Fig. 3.1C&2). In contrast, we expect PAHs formed in hot circumstellar envelopes or in the parent body to have $\Delta2\times^{13}$C close to zero (Fig. 3.1A&2).

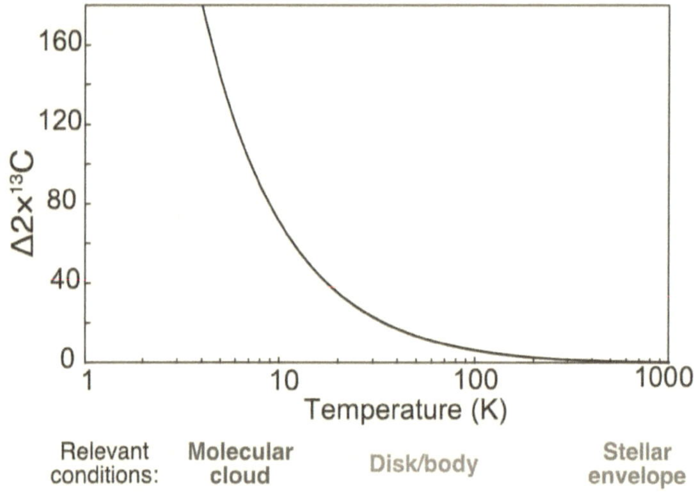

Figure 3.2: Predicted $\Delta2\times^{13}C$ values from our model.

The black curve shows our theoretical calculations of the $\Delta2\times^{13}C$ values expected for acetylene at different temperatures (See Supplemental Materials and Methods).Labels below the plot indicate the relevant environments at each temperature and use the same colors as Fig. 1.

Isotopic measurements of PAH samples

The Hayabusa2 spacecraft collected 5.4 g of material from two locations on Ryugu (Tachibana et al., 2022). Previous analysis of Ryugu samples has demonstrated that Ryugu is compositionally similar to Ivuna-type carbonaceous (CI) meteorites (Yokoyama et al., 2022). Soluble organic molecules previously have been extracted (Aponte, Dworkin, Glavin, et al., 2023; Naraoka et al., 2023) from Ryugu samples A0106 and C0107, which are aggregates of sub-millimeter grains collected from the first and second touchdown sites, respectively (Naraoka et al., 2023; Tachibana et al., 2022). Endogenous, complex soluble aromatic organics have been identified among the compounds extracted from these samples using the solvent dichloromethane (DCM) (Aponte, Dworkin, Glavin, et al., 2023). The molecules identified included several PAHs: the 2-ring naphthalene, 3-ring isomers phenanthrene and anthracene, and 4-ring isomers fluoranthene, and pyrene. The total PAH abundances in the Ryugu samples (8.6 to 32.4 nmol g^{-1}; Table 3.1; Fig. S3.2) were too low for isotopic analysis using traditional methods of gas-source isotope ratio mass spectrometry (IRMS) (Aponte, Dworkin, Glavin, et al., 2023; Naraoka et al., 2023).

We analyzed PAHs within the same DCM extracts of Ryugu as those previous studies (Aponte, Dworkin, Glavin, et al., 2023; Naraoka et al., 2023) using gas chromatography coupled with Orbitrap mass spectrometry (GC-Orbitrap)(Zeichner, Wilkes, et al., 2022). We measured the $\delta^{13}C_{VPBD}$ values of the five PAHs previously identified in Hayabusa2 sample A0106 and the $\delta^{13}C_{VPBD}$, δD_{VSMOW} and $\Delta2\times^{13}C$ values of the same five PAHs from C0107 (See Supplemental Materials and Methods). For comparison, we measured $\delta^{13}C_{VPBD}$ and $\Delta2\times^{13}C$ values of fluoranthene and pyrene extracted from the CC meteorite Murchison (See Supplemental Materials and Methods; Zeichner, Wilkes, et al., 2022). We also measured $\delta^{13}C_{VPBD}$ and $\Delta2\times^{13}C$ values of combusted plant biomass as a high temperature comparison(See Supplemental Materials and Methods). As a control sample, we used a serpentine blank processed in parallel with the Ryugu samples, and within this blank we detected no PAHs above the instrumental background (Table S1). Measured isotope values are listed in Table 3.1 and plotted in Fig 3.3.

PAH	Abund. (ppm)		$\delta^{13}C_{VPBD}\pm\sigma$ (n)			$\delta D_{VSMOW}\pm\sigma$ (n)			$2\times^{13}C/^{12}C\pm\sigma$ (n)			$\Delta2\times^{13}C\pm\sigma$	
	Ryugu A0106	Ryugu C0107	Terr. std.	Ryugu A0106	Ryugu C0107	Murch.	Terr. std.	Ryugu C0107	Terr. std.	Ryugu C0107	Murch.	Ryugu C0107	Murch.
Naph.	0.2	0.06	23.8 ± 0.5 (2)	27.5 ±24.6 (2)	20.3 ± 4.9 (3)	-	67.3 ± 0.4 (3)	7.2 ± 36.0 (4)	0.0048 ± 0.0060 (7)	0.0050 ± 0.0107 (7)	-	35±20	-
Phen.	0.33	0.04	24.4 ± 0.6 (2)	14.9 ± 6.9 (2)	12.6 ± 4.2 (3)	-	115.0 ± 1.1 (3)	458.9 ± 31.0 (4)	0.0095 ± 0.0046 (7)	0.0097 ± 0.0066 (7)	-	4±10	-
Anth.	0.15	0.04	24.1 ± 0.5 (2)	10.5 ± 0.5 (2)	16.7 ± 4.5 (3)	-	98.9 ± 0.3 (6)	581.3 ± 41.2 (4)	0.0095 ± 0.0043 (7)	0.0097 ± 0.0088 (7)	-	3±10	-
Fluor.	1.33	0.3	24.2 ± 0.4 (2)	27.1 ± 9.8 (2)	17.1 ± 2.5 (3)	9.9 ± 1.0 (3)	102.0 ± 1.2 (6)	137.0 ±18.9 (4)	0.0122 ± 0.0025 (7)	0.0125 ± 0.0050 (7)	0.0133 ± 0.008 (3)	9±5	51±13
Pyrene	6.19	1.13	25.2 ± 0.5 (2)	29.6 ±1.3 (2)	23.6 ± 2.1 (3)	11.3 ± 1.1 (3)	66.8 ± 1.2 (6)	68.0 ± 15.3 (4)	0.0122 ± 0.0023 (7)	0.0124 ± 0.0039 (7)	0.0126 ± 0.009 (7)	11±4	1±13

Table 3.1: PAH abundance (Abund., ppm) and isotope ratio measurements.

$\delta^{13}C_{VPBD}$, δD_{VSMOW}, $2\times^{13}C/^{12}C$ ratios and $\Delta2\times^{13}C$ values of PAHs extracted from Ryugu samples A0106 and C0107, the Murchison (Murch) meteorite, and terrestrial standards (Terr. Std.). $\delta^{13}C_{VPBD}$, δD_{VSMOW}, and $\Delta2\times^{13}C$ values are reported in units of ‰. The number of replicate analyses, n, is included in parentheses (21). σ is the standard error propagated in quadrature (See Supplementary Materials and Methods). All uncertainties on the abundances of PAHs in Ryugu samples are 25% relative and described in Supplementary Materials.

We found elevated $\Delta2\times^{13}C$ values of 9 to 51‰ (0.9 to 5.1%) for naphthalene, fluoranthene and pyrene from Ryugu, and fluoranthene from Murchison, which significantly exceed the expected stochastic values (all p-values < 0.005) (See Supplemental Materials and Methods). The $\delta^{13}C_{VPBD}$ values are the same (within analytical uncertainties) for PAHs with even and odd numbers of rings in Ryugu, and for fluoranthene and pyrene in Murchison. δD_{VSMOW} values for naphthalene, fluoranthene and pyrene from Ryugu are 320 to 570‰ higher than Ryugu's phenanthrene and anthracene.

Cold PAH formation processes

The $\Delta2\times^{13}C$ values we measured for naphthalene, fluoranthene and pyrene from Ryugu and fluoranthene from Murchison are larger than previously measured $\Delta2\times^{13}C$ values of ethane that was formed by processes at Earth-surface conditions ($\sim1‰$) (Clog et al., 2018). These large positive anomalies are consistent with the hypothesis that some (or all) of these PAHs were synthesized within cold (<50 K) environments (Figs. 3.1C, 3.3A&C), via a process that preferentially forms ^{13}C-^{13}C bonds. The clumped-^{13}C composition of all PAHs we measured is independent of molecular-average $\delta^{13}C_{VPBD}$ values, which span a narrow range. This is consistent with the low-temperature process that formed these PAHs having a high yield, such that the overall ^{13}C abundances of product PAHs approached those of the reactants. We would also expect PAHs formed within molecular clouds to have higher δD_{VSMOW} values than those of PAHs formed within other environments (Fig. 3.1C) (Sandford, 2002) for the same chemical physics reasons that cause interstellar CO or acetylene to be enriched in ^{13}C or ^{13}C-clumping, respectively.

The highest $\Delta2\times^{13}C$ value we measure, 51±13‰ for Murchison fluoranthene, is close to the acetylene equilibrium value at $\sim10K$, which we interpret as evidence that most of the fluoranthene in Murchison was synthesized in the ISM (Öberg, 2016). We suggest that low-temperature reactions were responsible for both the initial formation of C-C bonds and the subsequent assembly of high-molecular-weight aromatic compounds from smaller molecules, because otherwise the assembly of larger structures would dilute the initial clumping. The $\Delta2\times^{13}C$ value of 4-ring fluoranthene from Murchison is higher than those of the 2- and 4-ring PAHs from Ryugu, which have positive but lower $\Delta2\times^{13}C$ values.

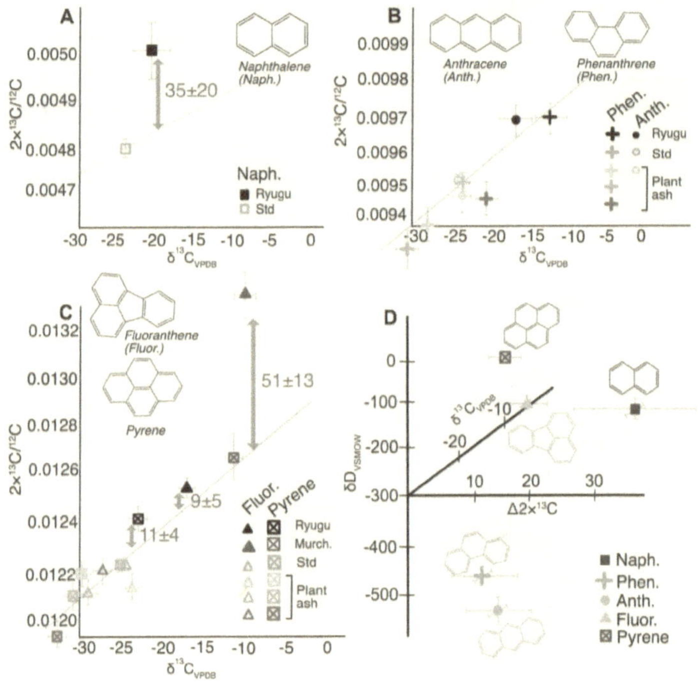

Figure 3.3: ^{13}C clumping measured from PAHs in the Ryugu samples.

$2\times{}^{13}C/{}^{12}C$ ratios are plotted as a function of $\delta^{13}C_{VPBD}$ for: (A) the 2-ring PAH naphthalene (black square); (B) 3-ring PAHs phenanthrene (black cross) and anthracene (black circle); and (C) 4-ring PAHs fluoranthene (black triangle) and pyrene (black open crossed square). Black symbols were measured from Ryugu sample C0107; terrestrial standards (Std) are shown with unfilled grey symbols. Panel C includes measured values for Murchison (Murch) meteorite (purple symbols). Panels B&C include residues of combusted plants (green symbols) (See Supplemental Materials and Methods, Karp et al., 2020). Predicted stochastic distributions are plotted as dotted grey lines (Supplemental Materials and Methods). Departures from stochasticity ($\Delta 2\times^{13}C\pm\sigma$ values) are indicated with labeled blue arrows (See Supplemental Materials and Methods). All error bars are 1σ and numerical values are listed in Tables 3.1&S3.3. (D) $\delta^{13}C_{VPBD}$, δD_{VSMOW}, and $\Delta 2\times^{13}C$ values for all five PAHs (see legend) in the Ryugu samples. Blue and orange ovals indicate PAHs that we interpret as formed by interstellar and circumstellar/parent body processes, respectively.

We suggest three scenarios for Ryugu's 2- and 4-ring PAHs: i) they could have formed at higher (but still low) temperatures than Murchison's fluoranthene, 20 to 50K. ii) they might be mixtures from two or more sources, with 20 to 50% formed in cold interstellar environments and the remainder formed (or reprocessed) at higher temperatures in other environments. iii) they could have formed from 2- or 3-carbon precursors with large positive $\Delta 2\times^{13}C$ values produced at low ISM temperatures, which were subsequently converted into larger aromatic molecules through reactions that diluted the initial $\Delta 2\times^{13}C$ signature.

Hot PAH formation processes

Three of the extraterrestrial PAHs that we examined have $\Delta 2\times^{13}C$ values that are consistent with calculated stochastic values (p-values > 0.15) (See Supplemental Materials and Methods): the 3-ring PAHs phenanthrene and anthracene from Ryugu and 4-ring PAH pyrene from Murchison. We do not expect synthesis of PAHs at high temperature to produce $\Delta 2\times^{13}C$ values that can be distinguished from zero using our methods (Supplemental Materials and Methods). We also measured no statistically significant (p-values > 0.04) deviations from the stochastic distributions of doubly-^{13}C-substituted species in the PAHs from combusted plant biomass (Fig. 3.3B&C; Table S3.3) (Supplemental Materials and Methods).

High-temperature processes that could have formed the 3-ring PAHs from Ryugu and pyrene from Murchison include: i) condensation in the outflows of carbon-rich AGB stars through bottom-up synthesis, such as the HACA mechanism, or ii) top-down catagenetic breakdown of larger carbonaceous dust grains. We expect that the formation of aromatic compounds by high temperature processes in circumstellar environments would produce PAHs with high $\delta^{13}C_{VPBD}$ (Abia et al., 2003) and low δD_{VSMOW} values, due to the nuclear fusion of deuterium lowering its abundance, and the subsequent convective mixing of D-poor ashes from the core of the star to the outer layers (Fig. 3.1A) (Busso, Gallino, and Wasserburg, 1999). This interpretation is consistent with our measurements of Ryugu phenanthrene and anthracene. A previous computational study of HACA formation of PAHs predicted it produces an excess of the 3-ring PAHs phenanthrene and anthracene (compared to other PAHs with different numbers of rings) (Kislov, Sadovnikov, and Mebel, 2013). Thus, PAHs formed by HACA could dominate the 3-ring species in Ryugu while making a small contribution to other PAH species.

Contributions from parent body processes

PAHs in Ryugu or Murchison samples that have $\Delta 2 \times^{13}C$ values close to zero could alternatively have been formed or altered by secondary processes (Fig. 3.1D). Previous experiments that performed pyrolysis (reactions that simulate catagenesis), on meteoritic IOM formed the same 2-, 3-, and 4-ring PAHs that we found in A0106 and C0107, at temperatures experienced by Ryugu's parent body ($<200°C$) (Vinogradoff et al., 2017). FTT synthesis is a less likely explanation for these samples, because FTT products such as alkanes are unlikely to undergo aromatization below 200°C (Yabuta et al., 2023).

We expect PAHs that formed on the parent body or experienced secondary processing to have low δD_{VSMOW} values (Fig. 3.1D). A previous experiment demonstrated that organic molecules formed under hydrothermal conditions from CO have δD_{VSMOW} values of -450 to $580‰$ (McCollom et al., 2010), which is consistent with the values that we measured for phenanthrene and anthracene from Ryugu (Fig. 3.3B). Hydrogen atoms in PAHs that were formed by primary interstellar processes could exchange isotopes and equilibrate with the parent body water (Alexander, Kagi, and Larcher, 1982). Ryugu is known to have experienced aqueous alteration (Yokoyama et al., 2022). Some CC meteorites are known to contain water that is depleted in D (including Murchison, which has δD_{VSMOW} $\sim$-440‰; Alexander, Kagi, and Larcher, 1982), but the δD_{VSMOW} value of Ryugu parent body water is unknown. Carbon-rich grains in Ryugu samples have heterogeneous δD_{VSMOW} values, ranging from 1000‰ (no D detected) to +10,000‰ (McCollom et al., 2010), indicating that parent body processes did not fully equilibrate hydrogen isotopic distributions in the Ryugu samples.

Formation of PAHs via synthesis in the parent body could explain the differences between the $\Delta 2 \times^{13}C$ values of fluoranthene and pyrene from Murchison. Murchison is known to have experienced aqueous alteration, which is variable even among different chips of the same specimen (Alexander, Bowden, Fogel, and Howard, 2015). Our measurements show that fluoranthene–which is the metastable isomer of the two four-ring PAHs that we studied (Wong and Westrum, 1971)— has $\Delta 2 \times^{13}C$ of 51‰, which is consistent with preferential ^{13}C-clumping at cold interstellar temperatures. Pyrene–the more stable isomer (Wong and Westrum, 1971)—has $\Delta 2 \times^{13}C$ values close to the stochastic distribution. Formation of PAHs under parent body conditions would be closer to thermodynamic equilibrium, favoring production of more stable isomers (pyrene preferred to fluoranthene) (Thiagarajan et al., 2020).

After mixing with pre-existing PAHs formed via other pathways, the additional stable PAHs would dilute and potentially erase any pre-existing interstellar isotopic fingerprints, such as ^{13}C clumping (see Supplementary Text).

Conclusions

We suggest that the diversity in $\delta^{13}C_{VPBD}$, δD_{VSMOW}, and $\Delta 2\times^{13}C$ values measured in PAHs from Ryugu and Murchison is consistent with PAHs being formed in at least two different settings (Supplemental Materials and Methods, Sephton, Watson, et al., 2015). Most of the 2- and 4-ring PAHs were synthesized in cold interstellar environments, while the PAHs we measured (including the 3-ring PAHs) were formed by synthesis or reprocessing within high- to moderate-temperature settings, such as circumstellar environments or the parent body. The non-zero ^{13}C clumping was measured in multiple PAHs from two samples that have different histories of alteration, collection, and (potentially) formation. By extrapolation, we infer that a large fraction of extraterrestrial PAHs were formed by interstellar processes.

Funding: S.S.Z. was funded by the NSF Graduate Research Fellowship. S.S.Z., A.E.H., and J.M.E. were funded by the NASA Emerging Worlds Grant ($18-EW18_2-0084$). S.S.Z., SBha, and J.M.E. were funded by the Simons Foundation Collaboration on the Origins of Life. J.M.E. was funded by the U.S. Department of Energy BES Grant ($DE-SC0016561$). J.C.A., J.P.D., D.P.G., E.T.P., and J.E.E were funded by the NASA Consortium for Hayabusa2 Analysis of Organic Solubles. H.Nar. and Y.Tak. were funded by the Japan Society for the Promotion of Science (21KK0062 to Y.Tak; 20H00202, JP20H00202, and JP20H05846 to H.Nar.). KGri was funded by the Australian Research Council through a Discovery Outstanding Research Award (DP130100577) and ARC Laureate fellowship (FL210100103).

Research data: Raw data from this study is available in an online repository, located at this DOI.

Author Contributions: Conceptualization: SSZ, JCA, JME, HNar, YTak, JPD; Methodology: SSZ, AEH, GDon, JME; Investigation: SSZ, JME, SBha, AEH, GDon, JCA, JPD, DPG, JEE, HVG, HNar, YTak, STac, ATK, AIH, KGri, KHF, HYur, TNak, TNog, ROka, HYab, KSak, TYad, MNis, ANak, AMiy, KYog, MAbe, TOka, TUsu, MYos, TSai, SatoshiTan, FTer, SNak, SWat, YTsu, KHam, KFuk, DAok, MHas, HMit, YChi, NOhk, NOO, SSak, ETP, HLM, F-RO-D, VVui, RThi, CWol, PSK, ARuf, TKog, JIsa, NHer, YOba, TosYos, DAra, HSug, AFur, KSas, HSat, YFur, JAok, KKano, SMN, TakYos, SatoruTan, MMor, MOno, FKab, KFuj, TYam, YKim; Visualization: SSZ, SBha; Data analysis: SSZ, AEH, GDon, SBha, JME; Modeling: SSZ, SBha, JME; Funding acquisition: SSZ, AEH, JME, JPD;

Project administration: SSZ, JME, JCA, JPD, STac, HNar, YTak; Supervision: JME, JPD; Writing – original draft: SSZ, SBha, JME; Writing – review editing: All authors

Data and materials availability: Summary processed isotope measurements are listed in Tables 1 and S3-S4. Raw experimental data are archived at Zenodo (50) in a form that complies with (ISAS data policies). Additional raw data and properties of the asteroid Ryugu are available in the Hayabusa2 Science Data Archives (DARTS). Samples of the Ryugu asteroid were allocated by the JAXA Astromaterials Science Research Group; the sample catalog is available online, and distribution for analysis is through Announcements of Opportunity available at https://jaxa-ryugu-sample-ao.net Samples from the Murchison meteorite were obtained from Clifford Matthew at the Field Museum in Chicago, IL (Zeichner, Wilkes, et al., 2022). The solvent extracted organic fractions of Ryugu and Murchison were fully consumed during the analyses in this study and in previous work (Zeichner, Wilkes, et al., 2022). The larger chip of Murchison from which the sample was taken remains at the Field Museum.

References

Abia, C. et al. (2003). "Understanding AGB carbon star nucleosynthesis from observations". In: *Publications of the Astronomical Society of Australia*. Vol. 20, pp. 314–323.

Abplanalp, M.J. and R.I. Kaiser (2020). "Implications for Extraterrestrial Hydrocarbon Chemistry: Analysis of Acetylene (C_2H_2) and D_2-acetylene (C_2D_2) Ices Exposed to Ionizing Radiation via Ultraviolet–Visible Spectroscopy, Infrared Spectroscopy, and Reflectron Time-of-flight Mass Spect". In: *Astrophys J* 889, 3.

Alexander, C.M.O.D., R. Bowden, M.L. Fogel, and K.T. Howard (2015). "Carbonate abundances and isotopic compositions in chondrites". In: *Meteorit Planet Sci* 50, pp. 810–833.

Alexander, C.M.O.D., R. Bowden, M.L. Fogel, K.T. Howard, et al. (2012). "The provenances of asteroids, and their contributions to the volatile inventories of the terrestrial planets". In: *Science* 337, pp. 721–723.

Alexander, R., R.I. Kagi, and A.V. Larcher (1982). "Clay catalysis of aromatic hydrogen-exchange reactions". In: *Geochim Cosmochim Acta* 46, pp. 219–222.

Allamandola, L.J., S.A. Sandford, and B. Wopenka (1987). "Interstellar Polycyclic Aromatic Hydrocarbons and Carbon in Interplanetary Dust Particles and Meteorites". In: *Science* 237, pp. 56–59.

Aponte, J.C., J.P. Dworkin, and J.E. Elsila (2015). "Indigenous aliphatic amines in the aqueously altered Orgueil meteorite". In: *Meteorit Planet Sci* 50, pp. 1733–1749.

Aponte, J.C., J.P. Dworkin, Daniel P. Glavin, et al. (2023). "PAHs, hydrocarbons, and dimethylsulfides in Asteroid Ryugu samples A0106 and C0107 and the Orgueil (CI1) meteorite". In: *Earth, Planets and Space* 75, p. 28.

Basile, B.P., B.S. Middleditch, and J. Or (1984). "Polycyclic aromatic hydrocarbons in the Murchison meteorite". In: *Org Geochem* 5, pp. 211–216.

Becke, A.D. (1993). "A new mixing of Hartree–Fock and local density-functional theories". In: *J Chem Phys* 98, pp. 1372–1377.

Bigeleisen, J. and M.G. Mayer (1947). "Calculation of equilibrium constants for isotopic exchange reactions". In: *J Chem Phys* 15, pp. 261–267.

Bigeleisen, J. and M. Wolfsberg (1957). "Theoretical and experimental aspects of isotope effects in chemical kinetics". In: *Adv Chem Phys* 1, pp. 15–76.

Brand, W.A. et al. (2014). "Assessment of international reference materials for isotope-ratio analysis (IUPAC technical report". In: *Pure and Applied Chemistry* 86, pp. 425–467.

Busso, M., R. Gallino, and G.J. Wasserburg (1999). "Nucleosynthesis in Asymptotic Giant Branch Stars: Relevance for Galactic Enrichment and Solar System Formation". In: *Annual Reviews in Astronomy and Astrophysics* 37, pp. 239–309.

Chiar, J.E. et al. (2013). "The structure, origin, and evolution of interstellar hydrocarbon grains". In: *Astrophysical Journal* 770.

Clog, M. et al. (2018). "A reconnaissance study of $^{13}C-^{13}C$ clumping in ethane from natural gas". In: *Geochim Cosmochim Acta* 223, pp. 229–244.

Cordiner, M.A. et al. (2019). "ALMA Autocorrelation Spectroscopy of Comets: The HCN/H^{13}CN Ratio in C/2012 S1 (ISON)". In: *Astrophys J Lett* 870, L26.

Doddipatla, S. et al. (2020). "Low-temperature gas-phase formation of indene in the interstellar medium". In: *Sciences Advances* 7.

Draine, B.T. (2016). "Graphite Revisited". In: *Astrophys Journal* 831, p. 109.

Eckstaedt, C.D.Vitzthum et al. (2012). "Compound specific carbon and hydrogen stable isotope analyses of volatile organic compounds in various emissions of combustion processes". In: *Chemosphere* 89, pp. 1407–1413.

Eiler, J.M. (2013). "The Isotopic Anatomies of Molecules and Minerals". In: *Annu Rev Earth Planet Sci* 41, pp. 411–441.

Eiler, J.M. et al. (2017). "Analysis of molecular isotopic structures at high precision and accuracy". In: *International Journal of Mass Spectrometry* 422, pp. 126–142.

Elsila, J.E. et al. (2012). "Compound-specific carbon, nitrogen, and hydrogen isotopic ratios for amino acids in CM and CR chondrites and their use in evaluating potential formation pathways". In: *Meteoritics and Planetary Science* 47.9, pp. 1517–1536. DOI: 10.1111/j.1945-5100.2012.01415.x.

Frenklach, M. and E.D. Feigelson (1989). "Formation of polycyclic aromatic hydrocarbons in circumstellar envelopes". In: *Astrophys J* 341, pp. 372–384.

Gilmour, I. and C.T. Pillinger (1994). "Isotopic compositions of individual polycyclic aromatic hydrocarbons from the Murchison meteorite". In: *Mon. Not. R. Astron. Soc* 269, pp. 235–250.

Glavin, D.P. et al. (2018). "The origin and evolution of organic matter in carbonaceous chondrites and links to their parent bodies". In: *Primitive Meteorites and Asteroids. Elsevier*. Ed. by N. Abreu. DOI: 10.1016/B978-0-12-813325-5.00003-3.

Graham, H.V. et al. (2022). "Deuterium Isotope Fractionation of Polycyclic Aromatic Hydrocarbons in Meteorites as an Indicator of Interstellar/Protosolar Processing History". In: *Life* 12, p. 1368.

Hofmann, A.E. et al. (2020). "Using Orbitrap mass spectrometry to assess the isotopic compositions of individual compounds in mixtures". In: *Int J Mass Spectrom* 457, p. 116410.

Huang, Y. et al. (2015). "Hydrogen and carbon isotopic ratios of polycyclic aromatic compounds in two CM2 carbonaceous chondrites and implications for prebiotic organic synthesis". In: *Earth Planet Sci Lett* 426, pp. 101–108.

Kaiser, R.I. and N. Hansen (2021). "An Aromatic Universe-A Physical Chemistry Perspective". In: *American Chemical Society Journal of Physical Chemistry A* 125, pp. 3826–3840.

Kaiser, R.I., D.S.N. Parker, and A.M. Mebel (2015). "Reaction dynamics in astrochemistry: Low-temperature pathways to polycyclic aromatic hydrocarbons in the interstellar medium". In: *Annu Rev Phys Chem* 66, pp. 43–67.

Karp, A.T. et al. (2020). "Fire distinguishers: Refined interpretations of polycyclic aromatic hydrocarbons for paleo-applications". In: *Geochim Cosmochim Acta* 289, pp. 93–113.

Kislov, V.V., A.I. Sadovnikov, and A.M. Mebel (2013). "Formation mechanism of polycyclic aromatic hydrocarbons beyond the second aromatic ring". In: *Journal of Physical Chemistry A* 117, pp. 4794–4816.

Lahuis, F. and E.F. Dishoeck (2000). "ISO-SWS spectroscopy of gas-phase C_2H_2 and HCN toward massive young stellar objects". In: *Astron. Astrophys* 355, pp. 699–712.

Lecasble, M. et al. (2022). "Polycyclic aromatic hydrocarbons in carbonaceous chondrites can be used as tracers of both pre-accretion and secondary processes". In: *Geochim Cosmochim Acta* 335, pp. 243–255.

Lyons, S.L. et al. (2020). "Organic matter from the Chicxulub crater exacerbated the K-Pg impact winter". In: *Proc Natl Acad Sci U S A* 117, pp. 25327–25334.

Manby, F.R. et al. (2019). *entos: A quantum molecular simulation package*. DOI: `10.26434/chemrxiv.7762646.v1`..

Martinez, O. et al. (2008). "Gas Phase Study of C^+ Reactions of Interstellar Relevance". In: *Astrophys J* 686, pp. 1486–1492.

Martins, Z. et al. (2015). "The amino acid and hydrocarbon contents of the paris meteorite: Insights into the most primitive cm chondrite". In: *Meteorit Planet Sci* 50, pp. 926–943.

Maruoka, T. et al. (2003). "Carbon isotope fractionation between graphite and diamond during shock experiments". In: *Meteorit Planet Sci* 38, pp. 1255–1262.

McCollom, T.M. et al. (2010). "The influence of carbon source on abiotic organic synthesis and carbon isotope fractionation under hydrothermal conditions". In: *Geochim Cosmochim Acta* 74, pp. 2717–2740.

McGuire, B.A., A.M. Burkhardt, et al. (2018). "Detection of the aromatic molecule benzonitrile (c-C6H5CN) in the interstellar medium". In: *Science* 359, pp. 202–205.

McGuire, B.A., R.A. Loomis, et al. (2021). "Detection of two interstellar polycyclic aromatic hydrocarbons via spectral matched filtering". In: *Science* 371, pp. 1265–1269.

Mebel, A.M., A. Landera, and R.I. Kaiser (2017). "Formation Mechanisms of Naphthalene and Indene: From the Interstellar Medium to Combustion Flames". In: *Journal of Physical Chemistry A* 121.5, pp. 901–926. DOI: `10.1021/acs.jpca.6b09735`.

Mehta, C. et al. (2018). "Caveats to exogenous organic delivery from ablation, dilution, and thermal degradation". In: *Life* 8.

Messenger, S.M. et al. (1998). "Indigenous polycyclic aromatic hydrocarbons in circumstellar graphite grains from primitive meteorites". In: *Astrophys J* 502, pp. 284–295.

Micelotta, E.R., A.P. Jones, and A.G.G.M. Tielens (2010a). "Polycyclic aromatic hydrocarbon processing in a hot gas". In: *Astron Astrophys* 510, A37.

– (2010b). "Polycyclic aromatic hydrocarbon processing in interstellar shocks". In: *Astron Astrophys* 510, pp. 1–19.

Naraoka, H. et al. (2023). "Soluble organic molecules in samples of the carbonaceous asteroid (162173) Ryugu". In: *Science* 379, eabn9033.

Naraoka, H., H. Mita, et al. (2002). "Shimoyama, dD of individual PAHs from the Murchison and an Antarctic carbonaceous chondrite". In: *Geochim. Cosmochim. Acta, Spec* Suppl. A 5.

Naraoka, H., A. Shimoyama, and K. Harada (2000). "Isotopic evidence from an Antarctic carbonaceous chondrite for two reaction pathways of extraterrestrial PAH formation". In: *Earth Planet Sci Lett* 184, pp. 1–7.

Naraoka, H., A. Shimoyama, M. Komiya, et al. (1988). "Hydrocarbons in the Yamato-791198 Carbonaceous Chondrite from Antarctica". In: *Chem Lett* 17.5, pp. 831–834.

Öberg, K.I. (2016). "Photochemistry and Astrochemistry: Photochemical Pathways to Interstellar Complex Organic Molecules". In: *Chem Rev* 116, pp. 9631–9663.

Pehr, K. et al. (2021). "Preservation and Distributions of Covalently Bound Polyaromatic Hydrocarbons in Ancient Biogenic Kerogens and Insoluble Organic Macromolecules". In: *Astrobiology* 21.

Pering, K. and C. Ponnamperuma (1971). "Aromatic Hydrocarbons in the Murchison Meteorite". In: *Science* 173, pp. 237–240.

Pizzarello, S. et al. (2001). "The organic content of the Tagish Lake meteorite". In: *Science* 293, pp. 2236–2239.

Plows, F.L. et al. (2003). "Evidence that polycyclic aromatic hydrocarbons in two carbonaceous chondrites predate parent-body formation". In: *Geochim Cosmochim Acta* 67, pp. 1429–1436.

RStudio (2020). "RStudio: Integrated Development for R". In: *RStudio, Inc.* p. Boston, MA.

Rustad, J.R. (2009). "Ab initio calculation of the carbon isotope signatures of amino acids". In: *Org Geochem* 40, pp. 720–723.

Sandford, S.A. (2002). "Interstellar processes leading to molecular deuterium enrichment and their detection". In: *Planetary and Space Science* 50, pp. 1145–1154.

Sears, D.W. (1975). "Temperature gradients in meteorites produced by heating during atmospheric passage". In: *Modern Geology* 5.

Sephton, M.A. (2002). "Organic compounds in carbonaceous meteorites". In: *Natural Product Reports* 19.3, pp. 292–311. DOI: 10.1039/b103775g.

Sephton, M.A. and I. Gilmour (2000). "Aromatic Moieties in Meteorites: Relics of Interstellar Grain Processes?" In: *Astrophys J* 540, pp. 588–591.

Sephton, M.A., C.T. Pillinger, and I. (1998). "Gilmour, 13C of free and macromolecular aromatic structures in the Murchison meteorite". In: *Geochim Cosmochim Acta* 62, pp. 1821–1828.

Sephton, M.A., J.S. Watson, et al. (2015). "Multiple Cosmic Sources for Meteorite Macromolecules?" In: *Astrobiology* 15, pp. 779–786.

Shingledecker, C.N. (2014). *Thermally induced chemistry of meteoritic complex organic molecules: a new heat-diffusion model for the atmospheric entry of meteorites*. arXiv preprint, 1–23.

Slavicinska, K. et al. (2022). "Link between Polycyclic Aromatic Hydrocarbon Size and Aqueous Alteration in Carbonaceous Chondrites Revealed by Laser Mass Spectrometry". In: *ACS Earth Space Chem* 6, pp. 1413–1428.

Smith, J.D.T. et al. (2007). "The Mid-Infrared Spectrum of Star-forming Galaxies: Global Properties of Polycyclic Aromatic Hydrocarbon Emission". In: *Astrophys J* 656, pp. 770–791.

Tachibana, S. et al. (2022). "Pebbles and sand on asteroid (162173) Ryugu: In situ observation and particles returned to Earth". In: *Science* 1016, pp. 1011–1016.

Thiagarajan, N. et al. (2020). "Isotopic evidence for quasi-equilibrium chemistry in thermally mature natural gases". In: *Proc Natl Acad Sci U S A* 117, pp. 3989–3995.

Tielens, A.G.G.M. (2008). "Interstellar polycyclic aromatic hydrocarbon molecules". In: *Annu Rev Astron Astrophys* 46, pp. 289–337.

– (2013). "The molecular universe". In: *Rev Mod Phys* 85, pp. 1021–1081.

Urey, H.C. (1947). "The Thermodynamic Properties of Isotopic Substances". In: *Journal of the Chemical Society (Resumed)*, pp. 562–581.

Vinogradoff, V. et al. (2017). "Paris vs. Murchison: Impact of hydrothermal alteration on organic matter in CM chondrites". In: *Geochim Cosmochim Acta* 212, pp. 234–252.

Wang, Z., E.A. Schauble, and J.M. Eiler (2004). "Equilibrium thermodynamics of multiply substituted isotopologues of molecular gases". In: *Geochim Cosmochim Acta* 68, pp. 4779–4797.

Wing, M.R. and J.L. Bada (1991). "Geochromatography on the parent body of the carbonaceous chondrite Ivuna". In: *Geochim Cosmochim Acta* 55, pp. 2937–2942.

Wong, W.-K. and E.F. Westrum (1971). "Thermodynamics of polynuclear aromatic molecules I. Heat capacities and enthalpies of fusion of pyrene, fluoranthene, and triphenylene". In: *Chem. Thermodynamics* 3, pp. 105–124.

Woon, D.E. and T.H. Dunning (1993). "Gaussian basis sets for use in correlated molecular calculations. III". In: *The atoms aluminum through argon. J Chem Phys* 98, pp. 1358–1371.

Yabuta, H. et al. (2023). "Macromolecular organic matter in samples of the asteroid (162173) Ryugu". In: *Science* 379, p. 9057.

Yada, T. et al. (2022). "Preliminary analysis of the Hayabusa2 samples returned from C-type asteroid Ryugu". In: *Nat Astron* 6, pp. 214–220.

Yokoyama, T. et al. (2022). "Samples returned from the asteroid Ryugu are similar to Ivuna-type carbonaceous meteorites". In: *Science* 33, p. 7850.

Zeichner, S.S., J.C. Aponte, et al. (2023). "Polycyclic aromatic hydrocarbons in samples of Ryugu formed in the interstellar medium". In: *Science*. DOI: `10.1126/science.adg6304`.

Zeichner, S.S., J. Nghiem, et al. (2021). "Early plant organics increased global terrestrial mud deposition through enhanced flocculation". In: *Science* 371.6528, pp. 526–529. DOI: `10.1126/science.abd0379`.

Zeichner, S.S., E.B. Wilkes, et al. (2022). "Methods and limitations of stable isotope measurements via direct elution of chromatographic peaks using gas chromotography-Orbitrap mass spectrometry". In: *Int J Mass Spectrom* 477, p. 116848.

Zolotov, M. and E. Shock (1999). "Abiotic synthesis of polycyclic aromatic hydrocarbons on Mars". In: *J Geophys Res Planets* 104, pp. 14033–14049.

Supplementary Materials and Methods

Ryugu and Murchison sample collection and preparation

The Hayabusa2 spacecraft collected 5.4 grams of sample from two locations on near-Earth carbonaceous asteroid (1623173) Ryugu and returned them to Earth under controlled conditions (Tachibana et al., 2022; Yada et al., 2022). The first sample (Chamber A) was collected 2019 February 21 from the surface and the second (Chamber C) 2019 July 11 from the near-subsurface exposed by the Small Carry-on Impactor (Tachibana et al., 2022). Unlike meteorites, the Ryugu samples have not been exposed to weathering on Earth (Yokoyama et al., 2022).

Samples were allocated by the Japanese Aerospace Exploration Agency (JAXA) for the analysis of organic molecules that can be extracted by solvents (Aponte, Dworkin, Glavin, et al., 2023; Naraoka et al., 2023). Samples allocated to these analyses included 17.15 mg of sample A0106, 17.36 mg of C0107, 17.56 mg of the Orgueil CI-type meteorite, and 16.21 mg of serpentine. Sequential solvent extraction began with hexane, followed by dichloromethane (DCM), then methanol, and finally water (Naraoka et al., 2023). The extracts of the serpentine sample served as a blank prepared alongside the Ryugu samples to evaluate whether the samples were being contaminated during the preparatory chemistry (Table S3.1).

200 μL of each solvent-extracted fraction (total of 800 μL) was sent to NASA Goddard Space Flight Center (GSFC) where untargeted gas chromatography (GC)-GC-time of flight (TOF)-mass spectrometry (MS) and GC-quadrupole MS analyses were performed (Aponte, Dworkin, Glavin, et al., 2023). These analyses identified organics that were endogenous to the Ryugu asteroid samples within samples A0106 and C0107 (Aponte, Dworkin, Glavin, et al., 2023). Endogenous organics were also found in the extracts of the Orgueil samples, but these samples were fully consumed by Aponte, Dworkin, Glavin, et al., 2023 during analyses. Following the analyses for Aponte, Dworkin, Glavin, et al., 2023, there were 5.5 μL and 23.6 μL remaining of the extracts of A0106 and C0107, respectively. A0106, C0107 and the serpentine blank samples (~50 μL remaining) were shipped in DCM solvent to Caltech in January 2022, where they were stored in a $-20°$C freezer until we performed the measurements for this study in April 2022.

~35mg of powder were removed from a chip of Murchison, a CM2 carbonaceous chondrite and 2.5mg were extracted by solvent for PAH analysis. The results of these analyses were originally included in Zeichner, Wilkes, et al., 2022 but were reanalyzed for this study. This sample was provided by Clifford Matthew's research group at the Field Museum, in Chicago, IL. Subsampling procedures followed procedures documented by prior studies that have extracted PAHs from meteorites for analysis (Aponte, Dworkin, and Elsila, 2015; Gilmour and Pillinger, 1994; Huang et al., 2015; Naraoka, Shimoyama, and Harada, 2000), and are described in full in Zeichner, Wilkes, et al., 2022.

Ryugu and Murchison were dissolved in different batches of solvent due to the different sample preparation histories. The Murchison measurement was performed in 2016 (Zeichner, Wilkes, et al., 2022). The Ryugu sample was sent within extract that had been added during solvent extraction at JAXA. Analyses of solvents and procedural blanks revealed that both solvents used during sample preparation at JAXA and at Caltech were free of PAHs, or there were none observable above background (Table S3.1).

Plant combustion sample collection and preparation

Dry material (a mix of leaves, duff, bark, twigs, and branches) from gingko (Ginkgo biloba), cycad (Cycas spp) and marri (Eucalyptus diversi) trees were combusted at 142°C, 594°C and 905°C, respectively, within a ventilated tent with an air blower to maintain sufficient O_2. Full details of the combustion experiments were described in Eckstaedt et al., 2012; Karp et al., 2020. The resulting burn residues were collected and stored in glass jars. Organic molecules were extracted via Soxhlet extraction with a 9:1 azeotrope of dichloromethane (DCM) and methanol (MeOH). The extracts that remained after previous measurements for Karp et al., 2020 were dried. Based on abundance analyses in Karp et al., 2020, we estimated that there were 100-1000s pmol of each PAH remaining per $2\mu L$ vial of sample. These vials were then sent from Curtin University to Caltech. Sealed vials of dried sample were stored in the same −20°C freezer as the Hayabusa2 sample extracts upon arrival. Combusted plant samples were resuspended in DCM prior to analysis on the GC-Orbitrap. We used the same batch of DCM across the measurements of combusted plant matter and standards, which we tested prior to addition for presence of any contaminant PAHs that could affect measurements. Plant extracts were measured on the GC-Orbitrap in September 2022, using the same methods (see below) as the Ryugu sample.

Standard sources and preparatory chemistry

Pure standards of naphthalene (Sigma Aldrich 84679-250 mg, Lot BCCCD4410), phenanthrene (Sigma Aldrich 73338-100mg, Lot BCCC1654), anthracene (Sigma Aldrich 31581-250mg, Lot BCCB5640), fluoranthene (Sigma Aldrich 11474-100mg, Lot BCCD1403) and pyrene (Sigma Aldrich 18868, Lot BCCB9571) were weighed into 2mL GC vials (~0.01mg uncertainty on standard mass). The PAH standards had known $\delta^{13}C_{VPDB}$ and δD_{VSMOW} values, based on molecular average isotope ratios measurements.

The molecular average $\delta^{13}C_{VPDB}$ values of the five PAH standards (one for each sample compound of interest) were characterized by elemental analyzer-isotope ratio mass spectrometry (EA-IRMS) at Caltech. The molecular average δD_{VSMOW} values of the PAH standards were characterized by temperature conversion elemental analysis (TC/EA) at The University of Wyoming Stable Isotope Facility. The mean $\delta^{13}C_{VPDB}$ and δD_{VSMOW} values are reported in Table 3.1. 1 mL of DCM (Honeywell GC 299-4 EB279-US) was added to each of the vials, which were then sonicated at room temperature for 10 minutes to ensure full dissolution of the PAH.

Overview of GC-Orbitrap mass spectrometry

Measurements were performed at Caltech on a Thermo Fisher QExactive mass spectrometer with samples introduced via a Trace 1310 GC equipped with a TG-5SilMS chromatographic column (30m long, 0.25 mm inner diameter, 0.25 μm film; see Eiler et al., 2017; Zeichner, Wilkes, et al., 2022 for more details regarding the GC-Orbitrap setup). We chose this column so that our measurements were consistent with previous studies measuring the isotopic compositions of extraterrestrial and terrestrial PAHs via GC-IRMS (Aponte, Dworkin, Glavin, et al., 2023; Karp et al., 2020; Lyons et al., 2020). The samples and standards were injected via a split-splitless injector (275° C) operating in splitless mode. Helium was used as a carrier gas at a constant flow rate of 1.4 mL min^{-1}. Oven ramps were based on a previous study (Aponte, Dworkin, Glavin, et al., 2023) but adjusted to separate the target compounds of interest: 50 to 110°C at 2°C min^{-1}, 5 minute hold at 110°C, 110 to 145°C by 2°C min^{-1}, 10 minute hold at 145°C, 145 to 170°C by 2°C min^{-1}, 5 minute hold at 170°C, 170 to 260°C by 2°C min^{-1}, 5 minute hold at 260°C, and 260 to 295°C by 20°C min^{-1}, followed by a 20 minute hold. A similar ramp was used to measure the fire products, but used a faster ramp from 50 to 110°C (20°C min^{-1}) to improve the separation between isomers.

Briefly, the GC-Orbitrap analyses were performed as follows. GC effluent was transferred directly into the ion source via a heated transfer line (260°C) where analytes were ionized via electron impact (EI; Thermo Scientific Extractabrite, 70eV). The ions were extracted from the source, subjected to collisional cooling during transfer through a bent flatpole, underwent mass-window selection using an Advanced Quadrupole Selector (AQS), then passed through an automatic gain control (AGC) gate prior to storage in the curved linear ion trap ('C-trap'), which is a potential-energy well generated by radio frequency and direct current potentials. Ions were accumulated in the C-trap until the total charge reached a user-defined threshold (the AGC target), then were introduced into the Orbitrap mass analyzer as a discrete packet. Within the mass analyzer, ions orbit between a central spindle-shaped electrode and two enclosing outer bell-shaped electrodes, moving harmonically at frequencies proportional to each ion's mass-to-charge ratio (m/z). The raw data product of this oscillation—referred to as the transient—was converted via fast Fourier transform into a data product that can be processed for isotope ratio analysis (see below).

We refer to each injection of a sample or standard as an acquisition. Each acquisition is comprised of several scans, and each scan is comprised of ion intensities (referred to within the Orbitrap software and within this supplementary text as "NL scores") and m/z ratios averaged by the Orbitrap over a short time interval. Scans are 200 ms long for measurements performed with a mass resolving power of 120k, and 350ms long for those with a mass resolving power of 180k. Mass resolving power for the Orbitrap is stated as the m/z of the mass spectral peak divided by the full width of that peak (Δm) at half of its maximum intensity ('FWHM'):

$$resolving\,power = \frac{(m/z)_a}{(m/z)_a - (m/z)_b} \tag{S3.1}$$

By convention, Orbitrap mass resolving power is reported for a peak where m = 200 Da, but functional resolving power varies as the square root of the inverse of peak mass; in other words, the mass resolving power is twice that of the reported value at 50 Da. See (31) for more details on the choice of resolving power for a given experiment.

Naphthalene, phenanthrene, anthracene, fluoranthene and pyrene do not fragment when introduced into the EI source. Therefore, the mass spectrum of each PAH consists of the unsubstituted molecular ion, and its singly- and doubly-substituted isotopologues (Fig. S3.1B&C). Mass resolving powers used were chosen to distinguish near-isobaric mass spectral peaks, most notably the ^{13}C- versus D-substituted isotopologues of the PAH of interest (Fig. S3.1B&C).

Every scan used for isotope ratio analysis was under AGC control, i.e., under conditions where the C-trap limited the number of ions entering into the Orbitrap in each scan. The 'AGC target' is the amount of ions that is allowed in when the AGC control is operational, and can be computed as the total ion current multiplied by the injection time over the course of a single acquisition. Ideally, the AGC target should not vary more than 10% relative between scans. We found that in some cases, the instrument was not under AGC control, which were identified as scans where the integration time reached the user-defined maximum threshold (3000 ms). In those cases, we rejected that measurement, and repeated the experiment either using an increased amount of sample or after adjusting other experimental parameters (see below).

Orbitrap analyses of PAH samples and standards

First, we performed experiments using standard mixtures. Each PAH standard solution was serially diluted and gravimetrically mixed with each other to form solutions of known molarities of the 5 PAHs of interest. We injected a solution that consisted of a mixture of 5 PAH standards, each present in concentrations of 5pmol μL-1. We prepared this standard mixture to have PAHs present in this initial concentration based on PAH abundances reported in samples A0106 and C0107 by a previous study (Aponte, Dworkin, Glavin, et al., 2023). To ensure complete mixing and homogenous concentration of the PAH standard within each vial, we pulled the solvent-PAH-mixture up into the syringe ten times (10×) each before filling a syringe with any standard-solvent mixture for dilution or mixing. Prior to analysis, we vortexed each sample for 20 seconds to re-suspend any compounds stuck to the walls in solution.

After injecting the mixture of standard PAHs, we injected 2 μL of A0106 and observed the chromatographic elution with an 8 Da mass window (Fig. S3.1A; experimental settings of 2×10^5 AGC target, and 120k resolving power, Eqn S3.1). The purpose of this first injection was compound identification and quantification of concentration relative to measured standards, as described above. The initial injection was used to identify potential co-eluents within the mass window of interest for the compounds, which could affect the measured isotope ratios (Hofmann et al., 2020). PAHs were present in concentrations of 0.05 to 1pmol per 2.5μL injection, corresponding to NL scores of 2×10^4 to 1×10^6. The chromatographic peaks of the PAHs were a small fraction of the total ion current for other co-eluting species in the mass spectrometric mass window 50 to 250 Da (Fig. S3.1A). In other words, when observed with a wide mass window, the PAHs of interest were not detectable above the sample background and Orbitrap instrumental background.

We compared the mass spectral peak intensities of the 5 PAH peaks in the first injection of the Ryugu sample to other standards with PAHs at concentrations of 0.01, 0.05., 0.1, 0.5, and 1pmol μL^{-1}. We used this first injection to calculate the absolute amounts of each PAH in the sample. We then mixed relative concentrations of PAH standards together that were scaled appropriately with the absolute and relative concentrations of PAHs within the sample. Following direct injection of the newly-mixed standard scaled to the concentrations of PAHs in A0106, we adjusted the amounts as necessary until the relative abundances matched those measured in A0106 (within a factor of 1.5×). This scaled standard mixture was used as a comparative standard for all subsequent injections and analyses of the Ryugu samples, which were measured in series to ensure close sample-standard matching in terms of concentration and experimental conditions. The relative abundances of PAHs in extracts of A0106 and C0107 were the same (Fig S3.2; although the absolute abundances in extracts of C0107 were about half those in A0106), so the same standard mixture was used as a reference for measurements of both samples. The same process was followed for mixing standards for comparison with the fire product samples.

We measured PAHs extracted from the Hayabusa2 samples in DCM using the direct elution method, which introduces the eluent directly from the GC column into the mass spectrometer without intermediate peak trapping in the GC oven (Zeichner, Wilkes, et al., 2022). This method allows for the characterization of multiple compounds within a complex mixture in a single experiment, which reduces the amount of sample needed compared to methods that capture and focus on a single compound in a chromatogram (Fig. S3.1) (Zeichner, Wilkes, et al., 2022). For the Ryugu samples, only small sample volumes were available and the concentrations of organics are low (picomole to sub-picomole of each PAH of interest per 1 μL of sample). The direct elution method allowed us to measure isotopic properties with per-mille-level precision, sufficient to distinguish the large isotopic differences between terrestrial and extraterrestrial organic molecules, and the differences in multiply-substituted chemical isotope effects between high and very low temperature conditions. Additional details regarding the direct elution method can be found in Zeichner, Wilkes, et al., 2022.

To reduce the effect of co-eluting compounds on the PAH ions, we used AQS to select for a narrower mass window (either 8 or 2 Da wide; see below) that were each centered around the mass of the PAH of interest and "jumped" between them over the course of the chromatographic elution (Fig. S3.1B&C). These mass jumps were timed to coincide with the elution time of each compound. Predicted time frames of elution for each PAH were constrained by observing the standard mixture of PAHs prior to analysis of the samples. This added mass window jump improved the shot noise limit (Eqn. S3.5) for each of our measurements of the isotopic composition of our standards by a factor of 2. The PAH isomers in this study elute close together in time (Fig. S3.1B&C) and have unsubstituted (and substituted) molecular ions of the same mass, allowing one mass window to be used for both compounds. The serpentine blank was run in series with the samples and standards; we found that PAH peaks were present at concentrations that were lower than the instrumental background (Table S3.1).

Abundance measurements

We determined the abundance of each PAH within each sample by comparing the measured NL score of the base peak of the molecular ion for that species with the NL score of the same peak measured for analysis of a standard of known amount. This calibration assumes a linear relationship between sample size and NL score. We made these assumptions due to the matched concentrations of each compound in samples and standards, and the negligibly small background peaks in procedural blanks run with pure DCM solvent (Tables 3.1, S3.1).

The abundances we measure differ from previous measurements of the same compounds in extracts of Ryugu samples and samples of the CI meteorite Orgueil (Fig. S3.2A) (Aponte, Dworkin, Glavin, et al., 2023), in both absolute and relative abundance. The concentration of naphthalene was much lower in our samples, which could be due to volatilization of naphthalene during sample transport and storage. We attempted to mitigate solvent loss by immediately recapping the vial each time it was punctured by the syringe for analysis.

For Murchison, we measured abundances that are one to two orders of magnitude higher than the abundances of PAHs in samples of Ryugu (Fig. S3.2A). We normalized the abundances of Ryugu PAHs to pyrene, and found these ratios are consistent with those in other CCs, though at the lower end of the previously measured range for naphthalene, phenanthrene and anthracene (Fig. S3.2B). Pyrene was the most abundant PAH in our Ryugu samples (Fig. S3.2B), which is consistent with the results of another study of the same sample extract (Aponte, Dworkin, Glavin, et al., 2023) as well as prior measurements for other CC meteorites (Fig. S3.2). Systematic uncertainties in PAH abundances could arise from each step of the preparation and quantification. We estimated the initial volume in the vial based on the number of replicates that were possible before there was no solvent left. However, some small amount of solvent could have volatilized (but we were unable to quantify this or include it in our uncertainty estimation). We estimated the uncertainty in the weighing of PAH standards to be 0.15 μg for each 1000 μg of standard weighed out. Each standard dilution had quantification uncertainties of about 5% relative, which we determined by doing replicate dilutions of the same solution and observing the intensity of that diluted standard on the GC Orbitrap. We increased this uncertainty by 3×, because each standard solution was diluted three times to arrive at the concentrations present within the standard mixture that were scaled to match the absolute and relative concentrations of A0106. The intensity of the mass spectral peaks on the Orbitrap varied by ~20% relative between replicate injections (which incorporates variation in the injection volume itself, heterogeneity in the concentration of the compound within the final standard mixture, and instrument variation). These three sources of uncertainty were combined in quadrature to determine the 25% relative uncertainty on the concentration calculations of each PAH, which accounts for most of the differences between our reported values and those of the previous study (Aponte, Dworkin, Glavin, et al., 2023).

Isotope ratio measurements

2.5 μL of Ryugu sample C0107 was injected for abundance and $\delta^{13}C_{VPDB}$ value measurements. Replicate acquisitions of 2.5 μL injections were performed with 8 Da mass windows for each compound (Fig. S3.1B). 4.5 μL of Ryugu sample C0107 was injected for δD_{VSMOW} value and $2\times^{13}C/^{12}C$ ratio measurements. Replicate acquisitions of 4.5 μL were performed with 2 Da mass windows (Fig. S3.1C). The combination of these two measurements were used to calculate standard-normalized ratios of D/H and $2\times^{13}C/^{12}C$, as described below. We performed all isotope ratio measurements under conditions where the total intensities of contaminant peaks were less than 20% of the intensity of the base peak of the target PAH (Fig. S3.1B&C), a threshold that was determined from previous work (Hofmann et al., 2020).

Each time we changed the mass window that the AQS was filtering for, we also varied the AGC targets and instrument mass resolving power, in order to optimize sensitivity, based on prior experience with similar measurements of standards. We chose the highest AGC target and the lowest mass resolving power possible for each mass window, where the full elution of the chromatographic peak of interest could be observed under AGC control. For the 8Da window measurement of the lower abundance compounds (naphthalene, phenanthrene and anthracene), we used a lower AGC target (2×10^4), whereas we used a 1×10^5 AGC target for the fluoranthene and pyrene mass window because those compounds were more abundant. All the 2 Da mass window measurements of the C0107 sample used an AGC target of 2×10^4 and a mass resolving power of 180k to ensure full separation of the ^{13}C and D mass spectral peaks.

Combustion-produced PAHs were measured at abundances of ~10pmol per injection (see Table S3.2 for injection volumes). 8 Da mass window measurements of combusted plant biomass were performed with 1×10^5 AGC target and 120k mass resolving power. 2 Da mass window measurements of combusted plant biomass were performed 2×10^4 AGC target and with mass resolving power of 120k. The 120k resolving power setting for combusted plant samples was not high enough to measure the δD_{VSMOW} values of those PAHs. A complete list of experiments and experimental parameters is provided in Table S3.2. Results are discussed further in the Supplemental Text.

Data processing for isotope ratio computation We adapted a previously described data analysis method (Sandford, 2002) to process the data from our measurements. The data were extracted and converted using the FT STATISTIC proprietary software (Thermo Fisher) designed to process Orbitrap data. The software was used to extract acquisition statistics from .RAW files, such as intensity and peak noise, and convert them into a .csv or .txt file. We then processed the data using a Python processing script (Zeichner, Nghiem, et al., 2021) Statistics and plots for our experiments were computed using MICROSOFT EXCEL and RSTUDIO (RStudio, 2020).

We manually selected data from the files that were processed by FT Statistic to choose time frames of elution for each of the chromatographic peaks. These time frames were consistent for each PAH and across acquisitions. In some cases, variations in the mass spectral peak intensities of each PAH led to different lengths of time that the compound was observed under AGC control; we chose a consistent window that encompassed the shortest time frame under AGC control for a given PAH across all replicate sample/standard acquisitions. This time frame was then used to integrate each peak for isotope analysis.

The Fourier transform of the transient data reports NL scores over time for selected-mass chromatograms, which were converted into ion counts:

$$N_{io} = \frac{S}{N} \times \frac{C_N}{z} \times \sqrt{\frac{R_N}{R}} \times \sqrt{\mu_s} \tag{S3.2}$$

where N_{io} is the number of observed ions, S is the reported signal intensity (NL score) for the molecular or fragment ion in question, N is the noise associated with that signal, R is the nominal mass resolving power (at m/z 200), R_N is a reference mass resolving power, and C_N, which is the number of charges corresponding to the noise at that reference resolving power (we adopted $C_N = 4.4$ from prior experiments; see Eiler et al., 2017 for additional details), z is the charge per ion as defined above, and μ_s is the number of microscans. Microscans are an optional subsampling that can occur within a single scan in an Orbitrap acquisition, and we always used 1 microscan.

We chose scans where the NL score was >5% of the maximum NL score of the chromatographic peak. NL scores from these chosen scans were converted into ions (Eqn. S3.2), and then the calculated number of ions for the substituted and unsubstituted mass spectral peaks were added for all chosen scans in the acquisition (Zeichner, Wilkes, et al., 2022). The sum of the substituted ions that were observed were divided by that of the unsubstituted ions to calculate the isotope ratio.

Due to the low abundance of ions observed within each mass window, the backgrounds for all compounds for both the 8 Da and 2 Da mass window experiments were not observed under AGC control (although the NL score of the background itself was observed). This meant that we could not calculate an isotope ratio for the background to apply a quantitative background correction to the isotope measurements. Instead, we report the observed intensities of the background alongside our reports of the maximum intensity of the eluting peak in Table S1. In all cases, the background never exceeded a fractional intensity of $\sim 10^{-3}$. Anthracene had the highest background relative to its peak height (Table S3.1), which we attribute to the near co-elution with phenanthrene, although we confirmed baseline separation of all five PAHs of interest from one another using the PAH standards injected at concentrations of 5 pmol per $1\mu L$.

We found that some measurements were performed with imperfect sample-standard intensity matching; in other words, the maximum NL scores of compounds varied between acquisitions. Because this variation also caused variation in the precision of the measurement performed during acquisitions, we weighted the acquisitions by the number of ions observed in each. Additional corrections were applied to process the measurements of A0106 (see below for more details). This was necessary because the amount of sample A0106 was limited ($5.5\mu L$) and half of it was used for the initial sample characterization described above. As a result, the two A0106 injections were not measured using the same method or with a direct comparison to a standard with matching concentrations of PAHs. All data for C0107 measurements were processed without need for these corrections for drift or scale, and so were processed as follows to compute error-weighted mean isotope ratios.

We computed error-weighted means of the sample-standard isotope ratio difference for multiple acquisitions of each PAH of interest. We weighted each sample and standard isotope ratio by its uncertainty to address the issue of having varied sample and standard intensities (above). Likewise, weighting was used to account for situations where we observed the same property with different mass window sizes (i.e., $2\times^{13}C/^{12}C$ measurements performed with both the 2 Da and 8 Da mass windows), which generally resulted in different errors for two measurements of the same molecular isotopic property. Error weighted means of the measured sample and standard isotope ratios were weighted by the shot noise limit (SNL), which is approximately equal to standard error (SE) of each acquisition (proportional to the number of analyte ions observed for each acquisition) as:

$$\hat{\mu} = \frac{\sum_{i=0}^{n} x_i}{\sigma_i^2} \times \frac{1}{\sum_{i=0}^{n} 1/\sigma_i^2} \tag{S3.3}$$

where x_i is the isotope ratio of each acquisition (n) and σ_i is the SNL of each acquisition i. The sample-standard difference was then computed by taking the ratio of the two error-weighted means; the resulting values are reported in Tables 3.1 and S3.3.

Reported uncertainties are 1 standard error (SE, σ), weighted as:

$$\sigma^2(\hat{\mu}) = \frac{1}{\sum_{i=0}^{n} 1/\sigma_i^2} \qquad (S3.4)$$

The SNL is a proxy of the relative standard error (RSE) of each acquisition, and also represents the number of ions that were observed of each compound for each acquisition (Eqn S3.2). SNL was used as the weight for both the calculation of $\hat{\mu}u$ and $\sigma(\hat{\mu})$ and was calculated as:

$$\frac{\sigma_{SNL}}{R} = \sqrt{\frac{1}{\sum C_{io}} + \frac{1}{\sum c_{io}}} \qquad (S3.5)$$

where $\frac{\sigma_{SNL}}{R}$ is the shot noise limit on the relative standard error for a single acquisition, $\sum C_{io}$ and $\sum c_{io}$, are the sums of the counts for the substituted and unsubstituted ions for the PAH of interest, respectively. Uncertainties in the sample-standard differences were propagated from the individual uncertainties calculated for sample and standard, respectively, and were added in quadrature. Isotope ratios for combusted plant material PAHs and their associated standard errors were computed without weighting, because there was enough sample to perform sufficient replicate measurements to reach the same levels of precision in each acquisition.

The measured sample-standard differences in isotope ratios were converted to standard reference scales as:

$$R_{ref} = \frac{R_{sample}}{R_{standard}} * R_{(TC)EA} \qquad (S3.6)$$

where R_{ref} is the isotope ratio of the PAH on the reference scales VPDB (for carbon) or VSMOW (for hydrogen), R_{sample} is the measured isotope ratio of PAH from the Ryugu extract or the burned plant biomass, $R_{standard}$ is the isotope ratio of the PAH standard measured the same conditions, and $R_{(TC)EA}$ is the isotope ratio of the standard on the reference scale, constrained by EA or TC/EA.

Isotope ratio analyses that were performed with an 8Da mass window measurement allowed for the number of substituted ions (i.e., ^{13}C) to be directly compared to the number of unsubstituted ions (i.e., ^{12}C), and therefore those measurements could be converted directly into the known international reference frame. Isotope ratios performed with 2Da wide mass windows (i.e., for δD_{VSMOW} or $2\times^{13}C/^{12}C$ ratio measurements) had to be converted into the base peak reference frame. In this case, we calculated the ion intensity ratio of the D or $2\times^{13}$C peak to that of the singly-substituted ^{13}C peak, and then we multiplied that value by the sample-standard difference in $^{13}C/^{12}C$ isotope ratio (after it had already weighted by $\frac{\sigma_{SNL}}{R}$). Finally, we multiplied that product by the accepted isotope ratio of the standard. Uncertainties for each ratio were combined in quadrature; error-weighted sigma values are reported for the average isotope ratio measurements acquired via GC-Orbitrap measurements, while standard deviations were reported for the molecular average isotope values acquired by EA and TC/EA.

To compute the doubly-substituted-^{13}C isotope ratios for the PAHs extracted from samples of Ryugu, we combined data that were acquired through measurements spanning both 8 Da and 2 Da mass windows. We converted 2 Da mass window acquisitions into the unsubstituted mass spectral peak reference frame as described above, then computed an error-weighted mean of the acquisitions from the two different types of measurements. In contrast, doubly-substituted ^{13}C isotope ratios measured for combusted plant-derived PAHs were based only on 2Da mass window measurements.

The singly-substituted ^{13}C and D isotope ratios are reported in δ notation, which presents the measured isotope ratio of a sample relative to the same ratio measured in a standard of known isotopic composition:

$$\delta = (\frac{R_{sample}}{R_{standard}} - 1) \tag{S3.7}$$

where R values are the error-weighted mean isotope ratios of the sample and the standard. We report singly-^{13}C and ^{2}H (D) substituted isotopic compositions. All $\delta^{13}C$ and δD values have been converted onto the VPDB and VSMOW international reference frame scales respectively, reported as $\delta^{13}C_{VPBD}$ and δD_{VSMOW} where the standard reference frames are indicated as subscripts. δ values are reported in units of per-mille (that is, 1000$\times$ the value calculated in 3.8, where 1‰ is equal to 0.1%)

(Brand et al., 2014).

We report doubly-^{13}C substituted isotopic compositions in two ways: i) as measured values of the $2\times^{13}$C/^{12}C ratio (standardized assuming the same ratio in the standard is stochastic); and ii) as $\Delta 2\times^{13}$C values, which are differences between the measured $2\times^{13}$C/^{12}C ratios and the expected $2\times^{13}$C/^{12}C ratios assuming a stochastic distribution of $2\times^{13}$C isotopologues for that molecule (see main text and Fig. 3.2). We calculated the $\Delta 2\times^{13}$C values using the following expression:

$$2\times^{13}C = \frac{\frac{2\times^{13}C/^{12}C_{sample-measured}}{2\times^{13}C/^{12}C_{std-measured}}}{2\times^{13}C/^{12}C_{std-expected}} - 1 \tag{S3.8}$$

In other words, we divided the 'measured' $2\times^{13}$C/^{12}C ratio of the sample by that of the standard, and then divide that quotient by the 'expected' $2\times^{13}$C/^{12}C ratio based on the stochastic distribution. Like δ values, $2\times^{13}$C/^{12}C values are reported in units of ‰.

Our calculation and interpretation of $\Delta 2\times^{13}$C values assumes that the standards have approximately stochastic proportions of isotopologues. We made this simplifying assumption for two reasons: i) Commercially available PAHs are synthesized at high temperatures, which we expect to produce $\Delta 2\times^{13}$C values close to zero (within ~1‰). ii) Previous measurements of $\Delta 2\times^{13}$C anomalies in natural and synthetic ethanes (Clog et al., 2018) are negligibly smaller than the precision of our experiments.

Sensitivity tests and potential errors due to differences in injection volume

Due to our limited amount of sample from Ryugu and low abundance of PAHs within that sample, in some cases we had to standardize measurements of our Ryugu samples to standards in a different volume of solvent (i.e., 1 μL and 5 μL). For two cases – naphthalene and phenanthrene – we found that replicate measurements of standards of that were injected with 1 μL and 5 μL solvent led to differences in their $2\times^{13}$C/^{12}C ratios (Table S3.5). In some cases, these differences were larger than the reported 1σ uncertainty of replicate measurements of C0107 with a 2 Da mass window. While it is possible that this variation is due to issues of experimental reproducibility, there is some possibility that different amounts of solvent would create different conditions in the source and thus isotopically fractionate the samples being measured. To account for the error that this difference may introduce, we combined the uncertainty reported for naphthalene and phenanthrene $\Delta 2\times^{13}C$ values in Table S3.5 in quadrature with the experimental uncertainty. Results of this error propagation are included in the error values reported in Table 3.1.

Measurements and standard corrections for acquisitions of A0106

There was insufficient sample to measure δD_{VSMOW} of A0106 PAHs. We attempted to measure $2\times^{13}C/^{12}C$ ratios of PAHs in A0106, but discarded the data because measurements of the $2\times^{13}C/^{12}C$ ratios for the standards bracketing the two acquisitions of the sample were not reproducible within $2\times$SE.

Uncertainties on measurements of $\delta^{13}C_{VPDB}$ values for PAHs from the Ryugu samples A0106 were consistently higher than other $\delta^{13}C_{VPDB}$ measurements reported in this study because there was only enough sample for two acquisitions, the first of which was a direct elution measurement without a mass window shift. Each acquisition was performed under different conditions with respect to the AGC target, and with imperfect sample-standard concentration matching during the iterative process described above. Differences in the isotope ratio of the naphthalene standard varied with different concentrations of naphthalene introduced into the instrument, perhaps because naphthalene is very volatile. Because of these compromising factors, we introduced additional corrections to the data processing procedure.

For sample-standard comparisons of A0106, we computed the isotope ratios of each sample-standard bracket directly (i.e., naphthalene, phenanthrene and anthracene measurements for separate acquisitions of sample A0106). In some cases, A0106 injections were measured in series with a standard that was not well matched in concentration (i.e., >2 orders of magnitude different in the case of some PAHs). In those cases, after measuring the sample, we measured two standards in series—one that was well-matched in its concentration of PAHs with those in Ryugu sample A0106, and one that was identical to the standard used for the initial sample-standard comparison. The difference in the isotope ratios of the two standards then was used to compute a constant, which reflected the shift in isotope ratio with sample size. This constant was used to correct the measured isotope ratio of the original standard measurement. Uncertainties for all ratios that were included in this conversion were propagated in quadrature, which led to much larger uncertainties (particularly for the $\delta^{13}C_{VPDB}$ value for naphthalene in A0106) than for measurements where sample-standard comparisons were performed more directly (i.e., for measurements of C0107).

Tests of statistical significance

We performed tests of statistical significance on each of our measured $\Delta 2 \times^{13}C$ values for PAHs in both Ryugu sample C0107 and combusted plant samples. The goal of these tests was to evaluate whether our measured values can reject the null hypothesis of a stochastic distribution ($\Delta 2 \times^{13}C = 0$). We tested null hypotheses H_0 against alternative two-sided hypotheses H_a to determine whether the mean of each measurement was significantly different from $\mu_0 = 0$. p-values were calculated based on Welch's one-sample, one-sided t-tests using the `t.test` function in RSTUDIO (RStudio, 2020).

As described in the main text, the naphthalene, fluoranthene and pyrene values are significantly different from $\mu_0 = 0$ (where $H_0 > H_a$, and p-values were 1.159×10^{-5}, 0.003, and 0.001, respectively). Phenanthrene and anthracene values were not significantly different from $\mu_0 = 0$ (where $H_0 = H_a$, and p-values were 0.16, and 0.45, respectively). $\Delta 2 \times^{13}C$ values of PAHs extracted from combusted plant samples were not significantly different from zero (p-values > 0.04; Table S3.3).

Reanalysis of Murchison data

Murchison data was adopted from a previous study (Zeichner, Wilkes, et al., 2022). The results reported in Zeichner, Wilkes, et al., 2022 included some data that were later determined to have been observed outside of AGC control (see above), so we re-processed the raw data and re-present the results in Table 3.1.

Murchison measurements were performed from 2016 December 2 to 6, on the same instrument that we used for measurements of Ryugu and combusted plant biomass samples. 1 to 10 pmol of each compound was measured per replicate injection, resulting in NL scores of 5×10^6 to 2×10^7 per analysis. Like most of our Ryugu PAH analyses (excepting the A0106 measurements described above), we only include sample-standard comparisons in our analyses where the sample and standards are concentration matched with one another within a factor of 2.

The Murchison data was processed in mostly the same way as the Ryugu data described above, with one difference: due to shifting elution times, each peak was integrated for a separate time window that corresponded to when the sample or standard acquisition was under AGC control (Table S3.2). Within that chosen time window, we selected scans where the NL score of the base peak was >5% of the maximum NL score that the base peak reached during its chromatographic elution.

Samples and standards were respectively averaged, and the quotient was taken to compute the ^{13}C/^{12}C ratio of the Murchison PAHs.

Murchison pyrene and fluoranthene $\delta^{13}C_{VPBD}$ values were converted onto the VPDB scale by multiplying the values by the ^{13}C/^{12}C ratio of the terrestrial standards measured in series with the Murchison PAHs. The $\delta^{13}C_{VPBD}$ values of the fluoranthene and pyrene standards were measured in triplicate at Caltech by EA, and were 25.7±0.5 and 25.7±0.7‰, respectively.

Theoretical calculations of acetylene equilibrium fractionation

Heavy isotope substitution slows vibrations in the fundamental modes of motion of a molecule (Fig. S3.4), which reduces the molecular vibrational energy and therefore the overall free energy of a molecule. This effect scales with the reduced masses of molecular motions. It is energetically favorable for molecules to arrange heavy rare isotopes (including ^{13}C) into bonds with each other, rather than distributing them randomly across all possible atomic sites. This preference for "nearest-neighbor" clumping also applies between separate molecules: it is energetically preferable for two molecules with C-C bonds to have one molecule with a ^{12}C-^{12}C bond and one with a ^{13}C-^{13}C bond, compared to two molecules with ^{13}C-^{12}C bonds. This preference for isotopic clumping is reduced at higher temperatures, due to the increasing effect of configurational entropy on overall free energy as temperature rises.

Acetylene (C_2H_2) is an abundant organic molecule in extraterrestrial environments, thought to be the precursor of organic molecules synthesized in many extraterrestrial environments (Kaiser and Hansen, 2021). The distribution of ^{13}C in acetylene molecules can be represented by the equilibrium reaction given by equation 3.1.

The relationship between vibrational frequency and molecular energy can be described though the partition function ratio between isotopically substituted and unsubstituted molecules Q' and Q, in the Bigeleisen-Mayer model (BM model) as described in Bigeleisen and Mayer, 1947; Urey, 1947:

$$\frac{Q'}{Q} = \frac{m'}{m}^{3r/2} \frac{\sigma_{sym}}{\sigma'_{sym}} \prod \frac{v'_i}{v_i} \times \frac{\exp -U'_k/2}{\exp -U_k/2} \times \frac{1 - \exp -U_k/2}{1 - \exp -U'_k/2} \tag{S3.9}$$

where m is the mass of the unsubstituted isotope exchanged, m' is the mass of the substituted isotope exchanged, r is the number of atoms of the element of interest in the molecule, $_{sym}$ and $'_{sym}$ are the symmetry numbers of the unsubstituted and substituted molecules, v and v' are the frequency of the bond vibration in the unsubstituted and substituted versions of the molecule, and U_k and U'_k are the energies for each vibrational fundamental mode k of the unsubstituted and substituted molecules. The fundamental vibrational modes of acetylene are shown in Fig. S3.4. U_k is:

$$U_k = \frac{h\nu_k}{k_b T} \tag{S3.10}$$

where h is Planck's constant, k_B is the Boltzmann constant, and T is temperature.

We used density functional theory (DFT) to calculate predictions for the optimal structure of the acetylene molecule and its fundamental vibrational modes (Fig. S3.4) in its unsubstituted, singly-^{13}C substituted and doubly-^{13}C substituted forms. These computations were performed for acetylene in vacuum. We used the B3LYP exchange correlation functional (Becke, 1993) and aug-cc pVTZ basis set (Woon and Dunning, 1993) in the ENTOS Qcore simulation package (Manby et al., 2019). Previous work has shown that the aug-cc-pVTZ basis set is sufficiently large for convergence of equilibrium energy and structures of small organic compounds (Rustad, 2009). The vibrational frequencies are then converted to reduced partition function ratios (RPFRs) using the BM model described in eqn S3.9 under the harmonic approximation.

From these RPFRs, we estimated $\Delta 2 \times^{13} C$ values:

$$\Delta 2 \times^{13} C \approx 1000 \times ln\left(\sqrt{\frac{Q''}{Q}}\Big/\frac{Q'}{Q}\right) \tag{S3.11}$$

where Q' and Q'' are the equilibrium partition functions for the singly and doubly-substituted isotopologues of acetyelene, and can be computed using Eqn S3.9 (substituting Q'' for Q' in the case of the double-substition). We find that the fractionation is negligible at high temperatures, and tens of per-mille at low temperatures (~10 K; Fig. 3.2).

We regard the DFT model of isotopic clumping in acetylene as a model of the maximum effect that could be produced by a reversible chemical process in a molecule containing C-C bonds. This is because the acetylene model i) involves two heavy isotopes sharing a bond as nearest-neighbors, and ii) that bond is a triple bond, which is considered a "high order" bond. Multiple substitutions separated by more bonds, and lower-order bonds tend to have weaker clumping effects (Wang, Schauble, and Eiler, 2004).

If the PAHs were not (exclusively) built from 2-carbon units, then acetylene might not be an appropriate model for large $\Delta2 \times^{13} C$ values we observe in Ryugu and Murchison PAHs. Nevertheless, we use acetylene as a model to assess the amplitudes of clumped isotope augment compound-specific isotope ratio measurements. anomalies that could arise through common chemical processes, even if the reactants and products involved in PAH formation were different. PAH formation might have involved irreversible reactions, which exhibit kinetic isotope effects rather than equilibrium isotope reactions, such as for acetylene (Eqn. 3.1 and Fig. 3.2). Chemical kinetic isotope effects have amplitudes similar to equilibrium isotope effects (Bigeleisen and Wolfsberg, 1957), so we regard this acetylene model as a way to provide order-of-magnitude estimates of the $2 \times^{13} C$ vibrational clumped isotope effects of PAHs and their variations with temperature. Predicted clumping for acetylene is the result of laws of chemical physics, which would also apply to other small two- and three-carbon precursors of PAHs.

PAH abundances

The abundances and distributions of PAHs vary among different petrologic types, between meteorites of the same type, and between different specimens of the same meteorite (Fig. S3.2B) (Slavicinska et al., 2022). These differences are not traceable to different extraction procedures (Basile, Middleditch, and Or, 1984; Pering and Ponnamperuma, 1971). Differences in abundance most likely reflect sample heterogeneity, not differences among meteorite classes, parent bodies or other larger-scale groupings, but could also reflect analysis by different techniques (GC-MS versus GC-Orbitrap, etc.). Thus, the differences in absolute and relative abundances of PAHs between the Ryugu samples and those found in CC meteorites potentially reflect differences that arise from removal or redistribution of PAHs over small spatial scales. These could be affected by PAH volatility and solubility during hydrothermal aqueous alteration on the parent body(Naraoka et al., 2023), geo-chromatographic separation (Slavicinska et al., 2022; Wing and Bada, 1991), or, in the case of meteorite samples, organic synthesis or decomposition of PAHs during passage through the Earth's atmosphere (Mehta et al., 2018; Sears, 1975;

Shingledecker, 2014).

Isotope values of PAHs

Molecular-averaged carbon isotope values ($\delta^{13}C_{VPBD} \pm 1\sigma$) for naphthalene, phenanthrene, anthracene, fluoranthene and pyrene from samples A0106 and C0107 from Ryugu are included in Table 1 and plotted in Fig. S3.3A. Deuterium isotope abundances ($\delta D_{VSMOW} \pm 1\sigma$) for naphthalene, phenanthrene, anthracene, fluoranthene and pyrene from C0107 are reported in Table 3.1 and plotted in Fig. S3.4B.

Our $\delta^{13}C_{VPBD}$ and δD_{VSMOW} values for Ryugu PAHs are within the range previously measured for meteorite samples (Fig. S3.3A). In our Ryugu samples, fluoranthene is ~8‰ enriched in ^{13}C relative to pyrene, which is similar to the enrichment previously observed in CCs (Gilmour and Pillinger, 1994; Huang et al., 2015; Naraoka, Mita, et al., 2002; Naraoka, Shimoyama, and Harada, 2000; Sephton and Gilmour, 2000; Sephton, Pillinger, and I., 1998). The $\delta^{13}C_{VPBD}$ values we measured for fluoranthene and pyrene from Murchison are consistent with past measurements of the same compounds using IRMS (Gilmour and Pillinger, 1994). δD_{VSMOW} values that we measured for Ryugu PAHs are all lower than the bulk δD_{VSMOW} values of both A0106 and C0107 measured in previous work, which were +252±13‰ and +269±13‰, respectively (Naraoka et al., 2023). $\delta^{13}C_{VPBD}$ values for plant-derived PAHs agreed (within 2σ with values previously measured by IRMS methods; Table S3.3; Karp et al., 2020).

We find that the 3-ring PAHs from the Hayabusa2 samples also have $\Delta 2 \times^{13} C \sim$ 0. This could be coincidental, but demonstrates differences between samples that we interpret as approximately stochastic (phenanthrene and anthracene) and not stochastic (naphthalene, pyrene and fluoranthene). Our finding that the combustion product PAHs have $\Delta 2 \times^{13} C$ values consistent with a stochastic distribution is as we expect based on our model of expected $\Delta 2 \times^{13} C$ values for PAHs that are formed by high temperature chemistries (Figs. 3.1&3.2).

Isotopic fractionation from mixing and other synthesis mechanisms

Mixing between two or more populations differing markedly in molecular-averaged $\delta^{13}C_{VPBD}$ values could result in positive $\Delta 2 \times^{13} C$ anomalies. Large ranges in $\delta^{13}C_{VPBD}$ (950 to +36,000‰) have been observed in prior studies of carbonaceous grains recovered from CC meteorites (Messenger et al., 1998), indicating that these meteorites are mixed aggregates of pre-solar and early-Solar-System materials differing in their stellar nucleosynthetic sources and post-formation processing. Mixing of two different formation processes with distinct isotopic signatures could produce positive $\Delta 2 \times^{13} C$ values. ^{13}C-clumping scales approximately with the square of the ^{13}C/^{12}C ratio, whereas mixing processes are approximately linear in plots of $2\times^{13}$C/^{12}C vs. $\delta^{13}C_{VPBD}$ (Eiler, 2013). Therefore, mixing between two pools of PAHs with very different $\delta^{13}C_{VPBD}$ values could also lead to differences in ^{13}C/^{12}C ratio that when mixed would deviate from the predicted curved stochastic distribution. This also means that the mixing phenomenon would require large ranges in $\delta^{13}C_{VPBD}$ values and correlations between the measured $\delta^{13}C_{VPBD}$ and $2\times^{13}$C/^{12}C ratios, which we do not find (Table 3.1, Fig. 3.3). We therefore hypothesize that the anomalies in multiple-^{13}C substitutions were generated by chemical or physical processes during the formation or modification of the PAHs.

Mixing could also occur between two or more populations that are similar in their molecular-averaged $\delta^{13}C_{VPBD}$ values but differ from one another in their $_{VSMOW}$ and $2\times^{13}$C contents. For example, if the proportions of PAHs generated by secondary processes differ from those generated via primary synthesis in the ISM, mixing could lead to some compounds (those formed in higher quantities by secondary processes) having near-stochastic $\Delta 2 \times^{13} C$ values, while other PAHs in the same sample (those formed in higher abundances in the ISM) retain large positive $\Delta 2 \times^{13} C$ anomalies. This mixing hypothesis is consistent with our PAH $_{VSMOW}$ measurements and their correlation with measured $\Delta 2 \times^{13} C$ values: The PAHs with non-stochastic $\Delta 2 \times^{13} C$ values have higher $_{VSMOW}$ values than the PAHs with $\Delta 2 \times^{13} C \sim 0$.

It is unknown whether the carbon atoms of PAHs can exchange, directly or indirectly, between PAHs and reactive carbon species under parent body conditions. However, formation of new PAHs from parent body substrates (water and DIC) would drive the $\delta^{13}C_{VPBD}$ values of the combined mixture of PAHs towards the $\delta^{13}C_{VPBD}$ value of the DIC. Murchison carbonate has been measured to have $\delta^{13}C_{VPBD} = +20$ to 80‰ (Sephton, 2002) (Fig. 3.1D)—values that are higher by 25 to 85‰ than the highest $\delta^{13}C_{VPBD}$ value we measured for PAHs in this study (Fig. 3.3A).

Alternatively, we consider it imaginable that the excesses in $\Delta 2 \times^{13} C$ values of naphthalene, pyrene and fluoranthene from Ryugu samples and fluoranthene from Murchison could arise from isotopic fractionation associated with energetic reprocessing in interstellar environments. However, only limited information is available on isotopic fractionation produced by shock wave chemistry in extraterrestrial environments; laboratory shock wave experiments producing diamond from graphite found no carbon isotope fractionations (Maruoka et al., 2003) but might not apply to PAHs.

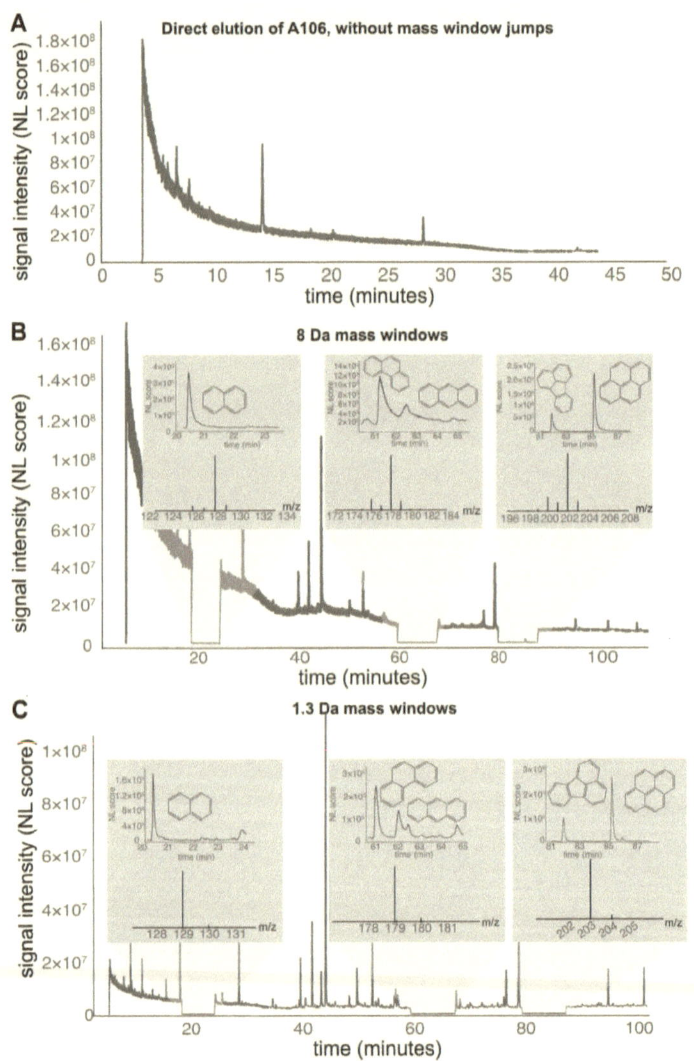

Fig. S3.1: Direct elution measurements of PAHs from Ryugu asteroid samples.

Caption continued on next page.

Figure S3.1: Direct elution measurements of PAHs from Ryugu asteroid samples.

After the initial injection of Ryugu sample A0106 into the Orbitrap MS, there were no chromatographic peaks visible for the PAHs of interest. All PAHs were present below the instrumental background (which is displayed here as the intensity, reported as an 'NL score,' of the 50-250Da mass window observed for the length of that samples acquisition in minutes). This first measurement was performed without mass window jumping. Subsequent measurements were performed by analyzing the compounds of interest using focused (B) 8 Da and (C) 2 Da windows. For each mass window, we display the 50-250Da mass window (the full zoomed out background figures). We also zoom in to each focused mass window, displayed as grey pop outs for each of the PAHs. Within each grey pop out, the top shows the chromatogram for the specific PAH of interest, and the bottom panel shows that PAHs mass spectrum. Each mass spectrum is normalized to 100%, with the maximum height being the intensity of the base peak of the PAH of interest. Full experimental conditions are reported in Table S3.2.

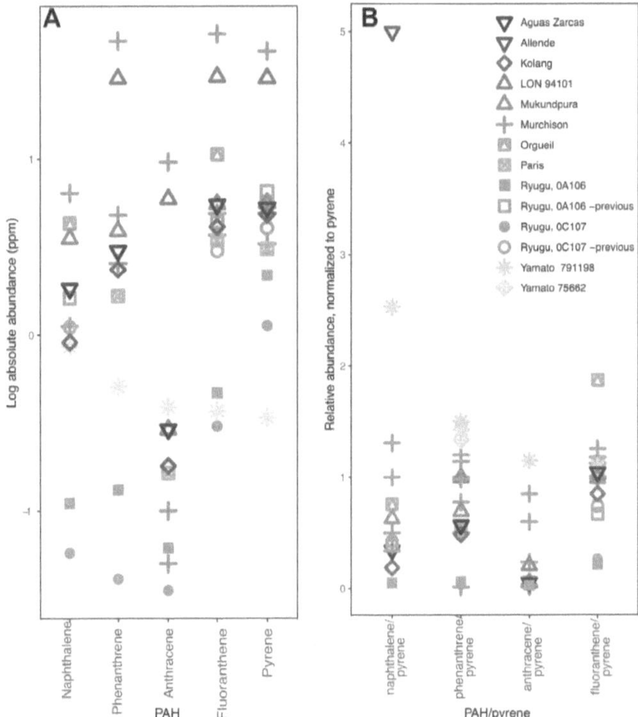

Fig. S3.2: Absolute and relative abundances of PAHs from meteorite and Ryugu samples.

(A) Log absolute abundances of 2-, 3-, and 4-ring PAHs in ppm (μg g^{-1}) extracted from Ryugu samples A0106 and C0107 measured in this study (filled cyan squares and circles) and in previous work (Aponte, Dworkin, Glavin, et al., 2023) (unfilled cyan squares and circles). For comparison, we show previously measured abundances of PAHs in extracts of other meteorite samples (grey symbols, meteorite names given in the legend). (B) The same data as panel A, but normalized to the abundance of pyrene. Aguas Zarcas, Mukundpura, and Kolang data are from Lecasble et al., 2022. Allende data are from Plows et al., 2003. LON 94101 data are from Huang et al., 2015. Murchison data are from Basile, Middleditch, and Or, 1984; Gilmour and Pillinger, 1994; Huang et al., 2015; Naraoka, Shimoyama, and Harada, 2000; Pering and Ponnamperuma, 1971; Wing and Bada, 1991. Orgueil data are from Aponte, Dworkin, Glavin, et al., 2023. Paris data are from Martins et al., 2015. Ryugu data are from this study and Aponte, Dworkin, Glavin, et al., 2023. Yamato 791198 are from Gilmour and Pillinger, 1994; Naraoka, Shimoyama, Komiya, et al., 1988. Yamato 75662 are from Gilmour and Pillinger, 1994.

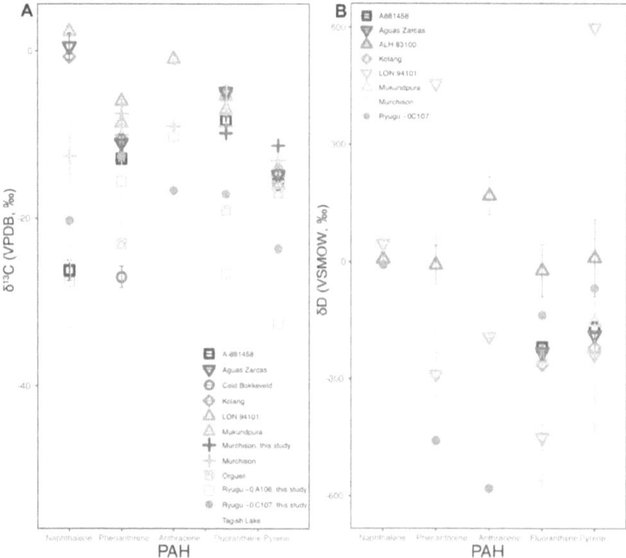

Fig. S3.3: $\delta^{13}C_{VPDB}$ and δD_{VSMOW} values of PAHs from meteorite and Ryugu samples.

(A) $\delta^{13}C_{VPDB}$ values measured for PAHs from samples of A0106 (unfilled cyan squares), C0107 (filled cyan circles), Murchison (purple plus signs), and other meteorites. A-881458 data are from Naraoka, Shimoyama, and Harada, 2000. Aguas Zarcas, Kolang, and Mukundpura data are from Lecasble et al., 2022. Cold Bokkeveld data are from Sephton and Gilmour, 2000. LON 94101 data are from Huang et al., 2015. Murchison data are from Gilmour and Pillinger, 1994; Huang et al., 2015; Naraoka, Mita, et al., 2002; Sephton, Pillinger, and I., 1998. Orgueil data are from (77). Ryugu data are from this study. Tagish Lake data are from Pizzarello et al., 2001. (B) δD_{VSMOW} values measured for PAHs from C0107 (filled cyan circles) and other meteorites. A881458 data are from Naraoka, Mita, et al., 2002. Aguas Zarcas, Kolang, and Mukundpura data are from Lecasble et al., 2022. ALH 83100 and LON 94101 data are from Graham et al., 2022. Murchison data are from Graham et al., 2022; Huang et al., 2015. Ryugu data are from this study.

Mode no.	Vibrational frequency, ν (cm⁻¹)		
	Unsubstituted	**Single ¹³C**	**Double ¹³C**
	Anti-symmetric C-H bending		
1	665.40	660.32	655.33
2	665.44	660.42	655.37
	Symmetric C-H bending		
3	768.43	767.34	766.15
4	768.49	767.40	766.21
	Symmetric C-C stretching		
5	2064.73	2032.36	1999.03
	Anti symmetric C-H stretching		
6	3408.25	3402.62	3398.16
	Symmetric C-H stretching		
7	3512.97	3500.56	3487.20

Fig. S3.4: Differences in the vibrational frequencies in the fundamental vibrational modes of acetylene.

Vibrational motions in acetylene can be divided into fundamental vibrational modes 1-7 above, where single or double isotopic substitutions of ¹³C change the vibrational frequency, ν, by the amount shown. This energy difference in the vibrational modes is the physical chemical phenomena that drives the clumped-¹³C anomaly (described further in the Main Text and in these supplemental materials and methods; Fig. 3.2). These motions were calculated using B3LYP/aug-cc-pVTZ level of DFT. Modes 1,2 and 3,4 are doubly degenerate.

Mass window	Compound	Ryugu A0106	Ryugu A0106 bkgd	Ryugu C0107	Ryugu C0107 bkgd	Serpentine blank
123.5 to 132.5	Naphthalene	5.29×10^5	4.49×10^3	2.60×10^5	3.75×10^3	3×10^3
173.5 to 182.5	Phenanthrene	2.95×10^5	3.63×10^3	6.68×10^4	3.70×10^3	6.42×10^2
	Anthracene	6.26×10^4	2.67×10^4	2.59×10^4	1.47×10^4	5.49×10^2
195.6 to 206.5	Fluoranthene	9.67×10^5	4.27×10^3	3.83×10^5	3.11×10^3	3.5×10^2
	Pyrene	3.30×10^6	7.62×10^3	1.25×10^6	5.22×10^3	3.45×10^2

Table S3.1: Measured background levels and blank control samples.

Intensity values ('NL scores') of the total ion current (TIC) for measurements of the serpentine blank are reported alongside the maximum NL scores of the PAH sample peaks of interest. NL scores for the PAH backgrounds (bkgd) in the measurements of Ryugu samples A0106 and C0107 refer to the intensity of the background for the 5 scans prior to the elution of the PAH's chromatographic peak.

Table S3.2: Experimental parameters used for each measurement of Ryugu, Murchison and combustion product samples.

PAH	Date	Sample	Run parameters			
			MW	TF	AGC	Res
Naph.	Apr-22	A0106–8Da ($2.8\mu L/1.1\mu L$)	124 to 132	20.7 to 22.15	2×10^4	120k
	Apr-22	C0107–8Da ($2.5\mu L/0.6\mu L$)	124 to 132	20.5 to 22.2	2×10^4	120k
	Apr-22	C0107–2 Da ($4.5\mu L/1\mu L$)	128.85 to 130.15	20.5 to 20.8	2×10^4	180k
	Sep-22	Marri ash–8Da ($1.1\mu L/1\mu L$)	—	—	—	—
	Sep-22	Marri ash–2Da ($1.1\mu L/1\mu L$)	—	—	—	—
	Sep-22	Cycads ash–8Da ($1\mu L/1.5\mu L$)	—	—	—	—
	Sep-22	Cycads ash–2Da ($1\mu L/1.5\mu L$)	—	—	—	—
	Sep-22	Gingko ash–8Da ($2\mu L/1\mu L$)	—	—	—	—
	Sep-22	Gingko ash–2Da ($2\mu L/1\mu L$)	—	—	—	—
	Dec-16	Murchison–8Da ($3\mu L/2\mu L$)	—	—	—	—

Continued on next page

Table S3.2: Experimental parameters used for each measurement of Ryugu, Murchison and combustion product samples. (Continued)

PAH	Date	Sample	Run parameters			
Phen.	Apr-22	A0106–8Da (2.8μL1.1μL)	174 to 182	61.1 to 62.1	2×10^4	120k
	Apr-22	C0107–8Da (2.5μL/0.6μL)	174 to 182	61.2 to 62.4	2×10^4	120k
	Apr-22	C0107–2 Da (4.5μL/1μL)	178.85 to 180.15	61.13 to 61.8	2×10^4	180k
	Sep-22	Marri ash–8Da (1.1μL/1μL)	174 to 182	36.2 to 37.2	1×10^5	120k
	Sep-22	Marri ash–2Da (1.1μL/1μL)	178.5 to 180.5	36.7 to 37.6	2×10^4	120k
	Sep-22	Cycads ash–8Da (1μL/1.5μL)	174 to 182	36.7 to 37.6	1×10^5	120k
	Sep-22	Cycads ash–2Da (1μL/1.5μL)	178.5 to 180.5	36.7 to 37.6	2×10^4	120k
	Sep-22	Gingko ash–8Da (2μL/1μL)	174 to 182	36.7 to 37.6	1×10^5	120k
	Sep-22	Gingko ash–2Da (2μL/1μL)	178.5 to 180.5	36.7 to 37.6	2×10^4	120k
	Dec-16	Murchison–8Da (3μL/2μL)	—	—	—	—
Anth.	Apr-22	A0106–8Da (2.8μL1.1μL)	174 to 182	62.1 to 63.35	2×10^4	120k

Continued on next page

Table S3.2: Experimental parameters used for each measurement of Ryugu, Murchison and combustion product samples. (Continued)

PAH	Date	Sample	Run parameters			
	Apr-22	C0107–8Da (2.5μL/0.6μL)	174 to 182	62.4 to 54	2×10^4	120k
	Apr-22	C0107–2 Da (4.5μL/1μL)	178	62.2 to 62.85	2×10^4	180k
			85 to 180.15			
	Sep-22	Marri ash–8Da (1.1μL/1μL)	—	—	—	—
	Sep-22	Marri ash–2Da (1.1μL/1μL)	—	—	—	—
	Sep-22	Cycads ash–8Da (1μL/1.5μL)	—	—	—	—
	Sep-22	Cycads ash–2Da (1μL/1.5μL)	—	—	—	—
	Sep-22	Gingko ash–8Da (2μL/1μL)	174 to 182	37.6 to 38.6	1×10^5	120k
	Sep-22	Gingko ash–2Da (2μL/1μL)	178.5 to 180.5	37.6 to 38.6	2×10^4	120k
	Dec-16	Murchison–8Da (3μL/2μL)				
Fluor.	Apr-22	A0106–8Da (2.8μL/1.1μL)	198 to 206	81.95 to 83.32	1×10^5	120k

Continued on next page

Table S3.2: Experimental parameters used for each measurement of Ryugu, Murchison and combustion product samples. (Continued)

PAH	Date	Sample	Run parameters			
	Apr-22	C0107–8Da (2.5μL/0.6μL)	198 to 206	81.95 to 83.32	1×10^5	120k
	Apr-22	C0107–2 Da (4.5μL/1μL)	202.85 to 204.15	82 to 83.05	2×10^4	180k
	Sep-22	Marri ash–8Da (1.1μL1μL)	198 to 206	55.85 to 57.5	1×10^5	120k
	Sep-22	Marri ash–2Da (1.1μL/1μL)	202.5 to 204.5	56.9 to 58	2×10^4	120k
	Sep-22	Cycads ash–8Da (1μL/1.5μL)	198 to 206	56.9 to 58	1×10^5	120k
	Sep-22	Cycads ash–2Da (1μL/1.5μL)	202.5 to 204.5	56.9 to 58	2×10^4	120k
	Sep-22	Gingko ash–8Da (2μL/1μL)	198 to 206	56.9 to 58	1×10^5	120k
	Sep-22	Gingko ash–2Da (2μL/1μL)	202.5 to 204.5	56.9 to 58	2×10^4	120k
	Dec-16	Murchison–8Da (3μL/2μL)	198 to 206	32.98 to 33.36 (Murch 2),	2×10^5	180k

Continued on next page

Table S3.2: Experimental parameters used for each measurement of Ryugu, Murchison and combustion product samples. (Continued)

PAH	Date	Sample	Run parameters			
				32.88 to 33.26 (Murch 3), 32.85 to 33.23 (Murch 4),32.86 to 33.24(Std 3),32.82 to 33.2 (Std 4)		
Pyrene	Apr-22	A0106–8Da $(2.8\mu L1.1\mu L)$	198 to 206	85.3 to 87	1×10^5	120k
	Apr-22	C0107–8Da $(2.5\mu L/0.6\mu L)$	198 to 206	85.3 to 87	1×10^5	120k
	Apr-22	C0107–2 Da $(4.5\mu L/1\mu L)$	202.85 to 204.15	85.25 to 87	2×10^4	180k
	Sep-22	Marri ash–8Da $(1.1\mu L/1\mu L)$	198 to 206	59.09 to 60.5	1×10^5	120k
	Sep-22	Marri ash–2Da $(1.1\mu L/1\mu L)$	202.5 to 204.5	60.2 to 61	2×10^4	120k
	Sep-22	Cycads ash–8Da $(1\mu L/1.5\mu L)$	198 to 206	60.2 to 61	1×10^5	120k
	Sep-22	Cycads ash–2Da $(1\mu L/1.5\mu L)$	202.5 to 204.5	60.2 to 61	2×10^4	120k
	Sep-22	Gingko ash–8Da $(2\mu L/1\mu L)$	198 to 206	60.2 to 61	1×10^5	120k

Continued on next page

Table S3.2: Experimental parameters used for each measurement of Ryugu, Murchison and combustion product samples. (Continued)

PAH	Date	Sample	Run parameters			
	Sep-22	Gingko ash–2Da (2μL/1μL)	202.5 to 204.5	60.2 to 61	2×10^4	120k
	Dec-16	Nurchison–8Da (3μL/2μL)	198 to 206	"34.5 to 32.93 (Murch 2), "	2×10^5	180k
				34.3 to 34.73 (Murch 3 + 4),34.37 to 34.8 (Std 3),34.33 to 34.76 (Std 4)		

Table S3.2: Experimental parameters used for each measurement of Ryugu, Murchison and combustion product samples.

Each row lists the parameters of a single measurement (meas.). For each measurement the "Sample" column indicates what it was, with what width of mass window it was measured, and what injection volumes were used for that sample and the standard it was compared to (i.e., Mass window (Injection volumes, sample/standard)). The remaining sample parameters for each PAH are listed in this order: mass window (MW; Da), time frame for peak integration (TF; minutes), AGC target (AGC), resolving power (Res). Standards were measured in series with samples, under the same experimental conditions and data processing parameters. Dashes indicate that that compound was not detected.

	T(°C)		phenanthrene			anthracene			fluoranthene			pyrene		
			$\delta^{13}C_{VPBD}$	$\Delta2\times^{13}C$	p-val	$\delta^{13}C_{VPBD}$	$\Delta2\times^{13}C$	p-val	$\delta^{13}C_{VPBD}$	$\Delta2\times^{13}C$	p-val	$\delta^{13}C_{VPBD}$	$\Delta2\times^{13}C$	p-val
gingko ash	142	this study	28.1±2.1	3.5±7.7	0.58	24.1±1.6	2.6±7.9	0.46	24.2±1.6	9.4±7.5	0.12	30.4±2.6	10.9±8.3	0.59
		previous work (49)	27.7±0.2	-	-	-	-	-	27.4±0.1	-	-	27.6±0.1	-	-
cycads ash	594	this study	31.8±2.7	4.4±8.5	0.22	-	-	-	29.3±2.7	2.5 ±7.3	0.28	31.7±2.7	4.5±7.4	0.68
		previous work (49)	26.8±0.1	-	-	-	-	-	28.1±0.1	-	-	26.±0.1	-	-
marri ash	905	this study	21.3±2.5	12.3±7.7	0.04	-	-	-	27.3±2.5	5.8±7.1	0.39	33.1±2.6	5.4±7.2	0.43
		previous work (49)	27.0±0.7	-	-	-	-	-	27.3±0.9	-	-	27.2±0.6	-	-

Table S3.3: Isotope measurements of combusted plant samples.

$\delta^{13}C_{VPDB}$ and $\Delta2\times^{13}C$ values of phenanthrene, anthracene, fluoranthene and pyrene measured from plant biomass that was burned at different temperatures. Both values are reported in units of ‰. $\delta^{13}C_{VPDB} \pm \sigma$ values were measured by the Orbitrap and are compared to previous measurements performed by GC-IRMS (49). Reported σ are 1SE. Reported p-values (p-val) are results from one-sided t-tests evaluating the significance of differences between measured $\Delta2\times^{13}C$ values and the null hypothesis of $\Delta2\times^{13}C = 0$.

	Standard $\delta^{13}C_{VPDB}\pm\sigma$(n) this study	Murchison $\delta^{13}C_{VPDB}\pm\sigma$(n) this study	Murchison $\delta^{13}C_{VPDB}\pm\sigma$ (52) from (52)	Murchison $\Delta2\times^{13}C\pm\sigma$ this study
Fluoranthene	25.69±0.46 (3)	9.9±1.10 (3)	5.9±1.1	51±13 (17)
Pyrene	25.63±0.66 (3)	11.3±1.1 (3)	13.1±1.3	1±13

Table S3.4: $\delta^{13}C_{VPDB}$ and $\Delta2\times^{13}C$ values of fluoranthene and pyrene from Murchison.

These results were obtained by re-processing of previously published data (Zeichner, Wilkes, et al., 2022). $\delta^{13}C_{VPDB}$ values from other previous work (Gilmour and Pillinger, 1994) listed for comparison, and additional prior measurements of Murchison $\delta^{13}C_{VPDB}$ values can be found in Fig. S3.3.

Compound	Injection Volume (mL)	Mass	Property measured	R value	Average	RSE
Naphthalene	5	128	^{13}C	0.107597	—	—
Naphthalene	1	128	^{13}C	0.10921	0.108404	0.0074
Naphthalene	5	128	$2\times^{13}$C	0.00486	—	—
Naphthalene	1	128	$2\times^{13}$C	0.005022	0.004941	0.0164
Anthracene	5	178	^{13}C	0.151578	—	—
Anthracene	1	178	^{13}C	0.151142	0.15136	0.0014
Anthracene	5	178	$2\times^{13}$C	0.010145	—	—
Anthracene	1	178	$2\times^{13}$C	0.010104	0.010124	0.002
Phenanthrene	5	178	^{13}C	0.151314	—	—
Phenanthrene	1	178	^{13}C	0.153305	0.15231	0.0065
Phenanthrene	5	178	$2\times^{13}$C	0.009989	—	—
Phenanthrene	1μL - jump method	178	$2\times^{13}$C	0.010128	0.010058	0.0069
Fluoranthene	5	202	^{13}C	0.169789	—	—
Fluoranthene	1	202	^{13}C	0.170018	0.169904	0.0007
Fluoranthene	5	202	$2\times^{13}$C	0.012612	—	—
Fluoranthene	1	202	$2\times^{13}$C	0.012706	0.012659	0.0037
Pyrene	5	202	^{13}C	0.169929	—	—
Pyrene	1	202	^{13}C	0.170139	0.170034	0.0006
Pyrene	5	202	$2\times^{13}$C	0.012681	—	—
Pyrene	1	202	$2\times^{13}$C	0.012793	0.012737	0.0044

Table S3.5: Sensitivity tests for injection volume.

Average isotope ratios and relative standard errors (RSE). RSEs are calculated by dividing the standard deviation by the average value. 'Property measured' refers to the isotopologue measured in reference to the unsubstituted base peak. Averages and RSEs are reported as differences between the 1 and 5μL injections and reported in the 1μL row.

POSITION-SPECIFIC CARBON ISOTOPES OF MURCHISON AMINO ACIDS ELUCIDATE EXTRATERRESTRIAL ABIOTIC ORGANIC SYNTHESIS NETWORKS

S.S. Zeichner, L.M Chimiak, et al. (2023). "Position-specific carbon isotopes of Murchison amino acids elucidate extraterrestrial abiotic organic synthesis networks". In: *Geochimica et Cosmochimica Acta*. DOI: 10.1016/j.gca.2023.06.010.

Sarah S. Zeichner[1], Laura Chimiak[1,2], Jamie E. Elsila[3], Alex L. Sessions[1], Jason P. Dworkin[3], José C. Aponte[3], John M. Eiler[1]

Affiliations: [1] Division of Geological & Planetary Sciences, California Institute of Technology, Pasadena, CA 90025, USA; [2] Department of Geological Sciences, UCB 399, University of Colorado, Boulder, CO 80309, USA; [3] Solar System Exploration Division, Code 691, NASA Goddard Space Flight Center, Greenbelt, MD, 20771. USA.

Introduction

Carbonaceous chondrites are amongst the most primitive materials in the Solar System, with heterogeneous compositions representing a diverse range of sources and formation processes spanning environments from the Interstellar Medium (ISM) to the Solar Nebula. Carbonaceous chondrites have up to 3 weight % organic carbon; studying these organic molecules can provide insights into prebiotic synthesis, potential sources of organic compounds on early Earth, and a framework for interpreting the origins of organic compounds that are the target of extraterrestrial sample return missions (Glavin, Alexander, et al., 2018). In particular, the Murchison meteorite is an ideal specimen for studying extraterrestrial organics. Murchison is a hydrated Mighei-type (CM2) meteorite that fell in 1969 near Murchison, Australia (Seargent, 1990), and has since been extensively studied due to its availability and relatively high concentration and chemical diversity of soluble organic compounds such as amino acids (review of studies in Glavin, Alexander, et al., 2018).

The abiotic synthesis of amino acids is of particular interest in origins of life research due to their biological ubiquity. Several pathways can synthesize amino acids abiotically; most require relatively complex series of reactions operating under specific conditions, imposing constraints on formation of these molecules on early Earth and the extraterrestrial settings where amino acids are found (Elsila, Aponte, et al., 2016; Glavin and Dworkin, 2009; Glavin, Elsila, et al., 2020; Pizzarello, Cooper, and Flynn, 2006). More than 90 amino acids have been identified in the Murchison meteorite, including proteinogenic ones (e.g., glycine, alanine, aspartic acid; Elsila, Aponte, et al., 2016; Glavin and Dworkin, 2009; Glavin, McLain, et al., 2020; Pizzarello, Cooper, and Flynn, 2006. The presence of non-proteinogenic amino acids, rare isotope enrichments in molecular average H-, N-, and C-compositions of meteoritic amino acids versus terrestrial biological amino acids (Table S4.5), and near-racemic mixtures of D/L enantiomers present compelling evidence that these amino acids are mostly endogenous to the Murchison meteorite (Burton et al., 2013; Elsila, Aponte, et al., 2016; Pizzarello, Huang, and Fuller, 2004).

A range of hypotheses exist regarding the reactions that could have synthesized $\alpha-$, $\beta-$, $\gamma-$, $\delta-$, etc. amino acids found in carbonaceous chondrites (summarized in Chimiak and Eiler, 2022; Elsila, Charnley, et al., 2012, among others): Fischer Tropsch/Haber Bosch-type synthesis (FTT-type; Botta and Bada, 2002; Burton et al., 2013; Hayatsu, Studier, and Anders, 1971; Kress and Tielens, 2001), Strecker-Cyanohydrin-type synthesis to form α-amino acids (Peltzer et al.,

1984), coupled with formaldehyde addition to form more complex soluble organic molecules (Chimiak and Eiler, 2022), ammonia involved formose reaction (Koga and Naraoka, 2017), reductive amination (Huber and Wächtershäuser, 2003), and Michael addition to form β-amino acids (Miller, 1957). Many of these processes require the presence of water and thus would likely occur in the meteorite parent body, but could form from precursors originally deriving from the ISM; amino acids could also potentially form in-situ within the protosolar nebula or ISM via UV irradiation of ice grains (Bernstein et al., 2002; Muñoz-Caro et al., 2002), although evidence for interstellar amino acids remains contested.

Structurally distinct groups of amino acids are expected to form via distinct pathways, so it seems likely that some combination of the preceding processes contributed to formation of the observed compounds (Burton et al., 2013; Elsila, Aponte, et al., 2016; Pizzarello, Huang, and Fuller, 2004; Sephton, 2002). However, it is unclear which process or processes dominate production of any specific amino acid, or whether different formation pathways occurred together, drawing on common substrates, or instead occurred independently from different pools of substrates.

Prior measurements of compound-specific carbon isotopic compositions of Murchison amino acids reveal many of them to be ^{13}C-enriched (Glavin and Dworkin, 2009; Pizzarello, Cooper, and Flynn, 2006; Sephton, 2002). For instance, α--alanine from the Murchison meteorite has been measured previously, yielding $\delta^{13}C_{VPDB}$ values of +25‰ to +52‰ (for either the D or the L enantiomer or a mixture of the two; Table S4.5; see Equation 4.1 for definition of δ). These values are higher than those typical of terrestrial biogenic alanine ($\delta^{13}C_{VPDB}$ usually < -20‰ but with values as high as -11‰ in C-4 plants; Cerling et al., 1997; Rasmussen and Hoffman, 2020). ^{13}C-enrichment suggests that synthesis of Murchison amino acids drew on precursors synthesized in the ISM. Interstellar molecules such as CO preferentially bond with heavier isotopes (i.e., ^{13}C; Charnley et al., 2004) relative to the population of free atoms, ions and radicals at cold temperatures ($\sim$10 K; Jørgensen et al., 2018; Öberg, 2016), a process driven by the greater thermodynamic stability of molecules that contain a heavy isotope. This process can be modeled by the following exothermic reaction — $^{12}CO + ^{13}C^+ \longleftrightarrow ^{13}CO + ^{12}C^+$ — where ^{13}C is thought to preferentially be sequestered within CO due to enhanced stability of the C-C bond at low temperatures. Interstellar hydrocarbons are formed from the reaction of C^+ with hydrogen, and are therefore thought to not be subject to such extreme stabilization of heavy-isotope substituted products, and thus are expected to be relatively ^{13}C-depleted.

However, it is challenging to make more specific interpretations of prior C isotope data of meteoritic amino acids. Molecular-average or "compound-specific" isotope analyses (CSIA) are limited in what they can reveal about mechanisms of amino acid synthesis, as the averaging across molecular sites can confound the effects of several variables (e.g., precursor isotopic compositions; reaction mechanisms; parent body processing; secondary alteration) on the isotopic composition.

Position-specific stable isotope analysis (PSIA) of meteoritic amino acids provides an opportunity to directly test abiotic synthesis hypotheses by tracing the isotopic compositions of putative precursor molecules to specific positions within reaction products. Prior studies have leveraged the potential of PSIA to understand reaction mechanisms for the formation of biological organics (e.g., Abelson and Hoering, 1961; DeNiro and Epstein, 1977; Melzer and O'Leary, 1987; Monson and Hayes, 1980), but until recently PSIA has required either chemical cleavage to isolate specific sites into molecules easily convertible into CO_2 for measurement via traditional isotope-ratio-mass-spectrometry methods, or NMR measurements of large samples (100's of mg) of pure analytes. The gas chromatography (GC)-Orbitrap is an emerging technology that enables precise, high-mass-resolution isotope measurements of relatively small samples ($\sim \mu$g and less) to record some features of position-specific isotopic variations in amino acids (and other molecules), where many compounds within a complex mixture can be introduced (following derivatization), subjected to an online separation and subsequently analyzed. The GC-Orbitrap platform enables measurements of position-specific isotopic composition by fragmenting analyte molecules in the ion source (Figure 4.1). Individual fragments sub-sample different combinations of molecular positions, such that several measured together may constrain differences in the rare isotope contents of different atomic sites of the parent molecule (Figure 4.2).

We previously applied this technique to constrain the carbon isotopic composition of α--alanine from Murchison using PSIA (Chimiak, Elsila, et al., 2021). We demonstrated that in the Murchison specimen analyzed, the high molecular-average $\delta^{13}C$ value (+26‰) was derived from a ^{13}C-enriched amine carbon (+142±20‰), and diluted by ^{13}C-depleted compositions for the carboxyl and methyl carbons (-29±10‰ and -36±20‰, respectively; Chimiak, Elsila, et al., 2021). Position-specific $\delta^{13}C$ values from Chimiak, Elsila, et al., 2021 suggest that Murchison α-alanine inherits the ^{13}C-enrichment in its amine carbon from a distinct, isotopically heavy carbon precursor (hypothesized to be carbonyl groups in aldehydes, which are suggested to derive from CO in the ISM), whereas the other two carbons are inherited from more ^{13}C-depleted precursors, either from ^{13}C-poor compounds in the ISM (e.g., hydrocarbons or HCN) or from carbon sources on the meteorite parent body (e.g., HCN; $\delta^{13}C_{VPDB}$ of KCN in Murchison = 5±3‰; Pizzarello, 2014; $\delta^{13}C_{VPDB}$ of cometary HCN = 16^{+262}_{-172}‰; Chimiak, Elsila, et al., 2021; Cordiner et al., 2019).

We argued that these precursors formed α-alanine (and other α-amino acids) via Strecker synthesis, plausibly in the meteorite parent body. However, this measurement and hypothesized reaction mechanism cannot provide constraints on how amino acids other than α-amino acids may have formed, and how the various classes of amino acids found in meteorites relate to one another. Performing PSIA on a several amino acids of distinct structural groups within the same sample presents an opportunity to relate position-specific isotopic compositions of amino acids, and their respective plausible formation pathways, to each other.

Here, we present measurements of position-specific carbon isotope values of α-alanine, aspartic acid, and β-alanine from Murchison and use our results to make specific assignments of the most plausible formation mechanisms for each of the measured amino acids. Many of these mechanisms have been introduced in prior studies (e.g., Elsila, Charnley, et al., 2012; Simkus, Aponte, Hilts, et al., 2019); we build upon these hypotheses with numerical constraints on isotope effects that serve to connect our position-specific isotope values with the isotopic compositions of potential precursors, thus building quantitative arguments in support of the most plausible reaction pathways. Our posited mechanisms offer a pathway for structurally distinct classes of amino acids to be synthesized from a small number of initial substrates, providing support for the hypothesis that a complex network of extraterrestrial organic synthesis created diverse amino acids from a common set of simple precursors. In particular, the doubly-carboxylated aspartic acid and the β-amino acid β-alanine have structural properties that differ from α-alanine and each other and cannot rely only on a Strecker synthesis mechanism as proposed for α-alanine's formation. PSIA allows us to distinguish the most plausible mechanisms for β-alanine and aspartic acid from a range of other mechanisms, in a way that cannot be resolved using CSIA alone.

Methods

Meteorite samples and amino acid standards

We analyzed two specimens of the Murchison meteorite – a 1.7g piece (Spring 2021) with some known terrestrial contamination and a 1.5g piece (Winter 2022) with evidence of minimal terrestrial contamination (Friedrich et al., 2019). The specimens were both acquired from the Field Museum of Natural History and previously described by Friedrich et al., 2019. Both samples were prepared via the same hydrolysis and desalting procedure at NASA Goddard Space Flight Center (GSFC) following the protocol from Elsila, Charnley, et al., 2012, along with a procedural blank prepared in parallel to confirm that no organics were being introduced during preparatory chemistry (Table S4.4).

Briefly, each sample was ground to a homogenized powder. The powder was split into two portions for more efficient extraction and then sealed in a glass ampoule with 1 mL of ultrahigh purity water (Millipore Integral 10 UV, 18.2 MX cm, <3 ppb total organic carbon) for 24 hours at 100°C. The water extract was separated, dried under vacuum, and hydrolyzed in 6 N HCl vapor (Tamapure-AA-10 purity, Tama Chemicals) for 3 hours at 150°C. This hydrolyzed extract was then desalted on a cation-exchange resin column (AG50W-X8, 100–200 mesh, hydrogen form, Bio-Rad), with the amino acids recovered by elution with 2 M NH_4OH (prepared from ultrahigh purity water and $NH_{3\,(g)}$ in vacuo) and dried under N_2. The Spring 2021 specimen was processed in this way in March 2021 and the Winter 2022 specimen in September of 2021 (Figure 4.1).

The abundances of amino acid enantiomers of both the Spring 2021 and Winter 2022 specimens were measured at GSFC via liquid chromatography with fluorescence detection and time-of-flight mass spectrometry (LC-FD/ToF- MS) using methods described in Glavin, Callahan, et al., 2010. Following LC-FD/ToF-MS analysis, the extracts were sent separately to Caltech along with a GSFC analytical blank processed in parallel. Upon arrival at Caltech, both the Murchison extract and the GSFC analytical blank were stored in a freezer until analyses in April 2021 and January 2022, respectively.

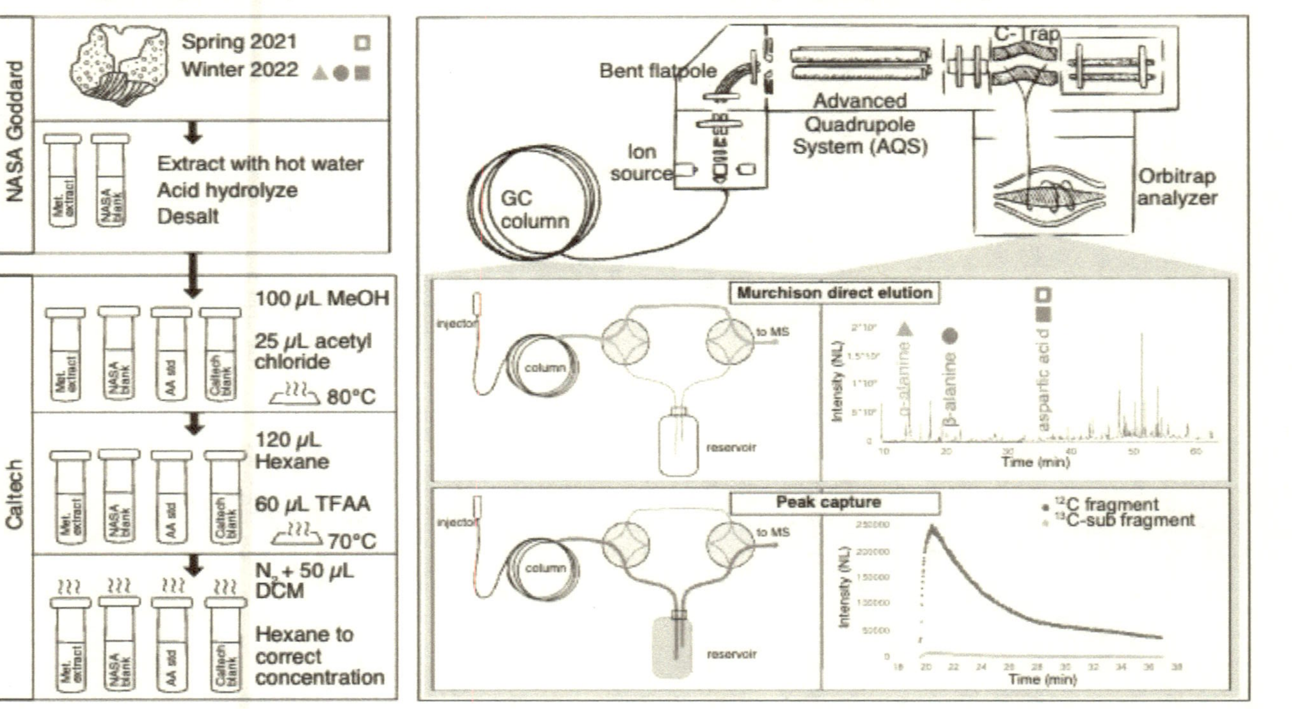

Figure 4.1: Preparatory chemistry and Orbitrap mass spectrometry.

Caption continued on next page.

Figure 4.1: Preparatory chemistry and Orbitrap mass spectrometry.

(left, Preparatory chemistry) Meteorite (Met.) samples were extracted, hydrolyzed, and desalted at NASA Goddard, before being sent to Caltech and derivatized in parallel with amino acid standards (AA std) and procedural blanks for introduction into the GC Orbitrap. (right, GC-Orbitrap mass spectrometry) GC-Orbitrap instrumental setup. Samples are introduced via gas chromatography into the mass spectrometer, where compounds are ionized, accelerated through the bent-flatpole, and mass-separated in the quadrupole before being stored in the C-trap, and injected into the Orbitrap. Two valves within the GC are used to direct the effluent through an internal plumbing that enable two types of measurements: direct elution and peak capture. Direct elution measurements allow the user to optimize chromatography and identify peak elution timing (Zeichner et al., 2022), while peak capture measurements capture a single fragment of interest within a stainless-steel reservoir and allow it to slowly purge into the MS over time.

For comparison to amino acids within meteorite extracts, we prepared pure amino acid standards of aspartic acid, β-alanine, and α-alanine, whose molecular average carbon isotopic composition ($\delta^{13}C_{VPDB}\pm 1\sigma_{SE}$) were constrained by Elemental Analyzer-Isotope Ratio Mass Spectrometry at Caltech to be -22.14 ± 0.07‰ (n = 3), -27.09 ± 0.40‰ (n = 3) and -19.60 ± 0.24‰ (n=3), respectively (Table S4.1). In addition to the pure extracts, we mixed unlabeled standards with a pure ^{13}C label at each site (10% label by mass; Supplemental Materials and Methods). This labeling experiment allowed us to identify fragments for each amino acid that sample a distinct combination of carbon positions to constrain each site, which can be converted into position-specific measurements (Figures 4.2&4.6).

Preparatory chemistry

Prior to analyses, meteorite extracts were redissolved in 1 mL of 3:1 H_2O:MeOH. After they were allowed to stand for 20 minutes, we sonicated them for 10 minutes to ensure full dissolution and homogenization within the sample vial. The solutions were transferred to 2 μL GC vials for derivatization chemistry. Half of the Spring 2021 sample was derivatized for analysis, and the full Winter 2022 sample was derivatized. To ensure quantitative transfer of sample when the full sample was being used, sample vials were rinsed three times with 50 μL of MilliQ water, which was also transferred into the GC vial. At this stage, a second blank was introduced to trace any potential sources of contamination during the derivatization chemistry. Samples, standards, and blanks were dried down fully under N_2 and derivatized in parallel as follows (Figure 4.1).

For analysis on the gas chromatograph (GC), we prepared derivatives by methylating the carboxyl group and trifluoroacetylating the hydroxyl and amine groups, producing N,O-bis(trifluoroacetyl) methyl esters (monoisotopic masses 257.05 Da, 199.05 Da, and 199.05 Da for aspartic acid, β-alanine, and α-alanine, respectively; Figure 4.1). The derivatization protocols were adapted from Corr and others and performed at Caltech (Corr et al., 2007). The samples and standards within GC vials were dissolved in 100 μL of anhydrous methanol (Sigma Lots SHBK8757 and 8HBN2144 for the Spring 2021 and Winter 2022 samples, respectively). Standard vials were placed on ice, where 25 μL of acetyl chloride (Sigma Lots BCBW8067 and S6039952B0 for the Spring 2021 and Winter 2022 samples, respectively) was added dropwise. Vials were heated at 80°C for 1 hour, then gently dried under N_2. Samples were dissolved in 120 μL of hexane, and 60 μL of trifluoroacetic anhydride (TFAA; Sigma Lots SHBK5942 and S7307761110 for the Spring 2021 and Winter 2022 samples, respectively) added to the vials and heated at 110°C for 30 minutes, after which they were briefly dried under N_2 until 25 μL of solvent remained in the vial. 100 μL of DCM was added to the vials to aide in eliminating residual derivatizing agent, then left to dry under N_2 until 10 μL were left. Derivatized standards were suspended in hexane (50 μL for samples and 1000 μL for standards). Standards were serially diluted to reach appropriate concentrations for introduction into the GC Orbitrap (i.e., 10s-100s pmols per 1 μL injection). Blanks were diluted in hexane to reach the same concentration of the meteorite samples. Samples, standards, and procedural blanks were derivatized in parallel but dried down separately to avoid cross contamination (Table S4.4).

Isotope analysis via GC-Orbitrap

Isotope analyses of amino acid mass spectral fragments were performed on a Q-Exactive Orbitrap mass spectrometer with samples introduced via a Trace 1310 GC equipped with a split/splitless injector operated in splitless mode with a TG-5SIL MS column (30 m (Spring 21) or 60m (Winter 22), 0.25 mm ID, 0.25 μm film thickness) and a 1 m fused silica capillary pre-column (220 μm /363 μm VSD tubing; Figure 4.1). Prior to position specific analyses of each amino acid, we performed direct elution measurements of samples and standards to confirm oven ramps, elution timing of each compound of interest, and detect any potential extraneous ions that co-eluted with analyte fragments within the mass window of interest (Figure 4.1 & S4.1; Supplemental Materials and Methods; Zeichner et al., 2022). For the Spring 2021 sample, the following oven program was used: 50 to

85° C by 10°C per minute, 85 to 100°C by 1°C per minute with a 9 minute hold, and 100 to 180°C by 4°C per minute with a 10 minute hold. For the Winter 2022 sample, the following oven program was used: 50 to 85°C by 10°C per minute, 85 to 100°C by 1°C per minute with a 9 minute hold, and 100 to 180°C by 4°C per minute with a 58 minute hold. Special care was taken to optimize chromatography to ensure that amino acids of interest did not co-elute with nearby compounds, and that prior compound tails did not contribute mass fragments at abundances above baseline within the mass range for each measurement.

Following the optimization of GC conditions, we performed peak capture measurements using a Silco-coated stainless steel reservoir (10 cc; SilcoTek, Bellefonte, PA; Figure 4.1; Chimiak, Elsila, et al., 2021; Eiler et al., 2017; Wilkes et al., 2022). We used valves to route the eluent into the reservoir during the elution time frame of the amino acid of interest (Figure 4.1), thus capturing the compound such that we could slowly purge it into the Orbitrap to observe the fragments of interest for a longer period of time than traditional GC peaks (~ minutes to tens of minutes instead of 10s of seconds).

We performed Orbitrap measurements of the carbon isotope ratios of fragment ions of the derivatized amino acids by focusing the Advanced Quadrupole System (AQS) of the QExactive platform to mass windows 6.5 or 7 Da wide, centered around the mass of the unsubstituted mass spectral peak of each fragment. Focused mass window measurements reduce the ions observed for a given acquisition so that each measurement is dominated by the fragment ions of interest, reducing the contribution of background and contaminant ions (Supplemental Materials and Methods). In some cases, narrowing the mass window was not able to eliminate all contaminants, like for some of the aspartic acid mass spectral fragments in the Winter 2022 sample, as the aspartic acid abundances were so low (Table 4.1) and the mass spectral peaks in some cases were less than one order of magnitude above the background. In the case of one aspartic acid fragment within the Winter 2022 sample (198 Da), we were not confident that we would be able to perform a measurement free from artifacts introduced by contaminant ions. In general, we only included measurements of mass spectral fragments where contaminant ions made up <20% of the total ion count (TIC; Supplemental Materials and methods), as suggested by Hofmann et al., 2020 in a prior study investigating the effects of contaminant ions on isotope ratios of other compounds in the same mixture (Hofmann et al., 2020). Hofmann et al., 2020 also demonstrated that resolutions of 60k mitigate the effects of contaminant ions (i.e., those that make up less than <20% of the TIC but that are still prominent in the mass window of interest) on the isotope ratios of the compound of interest; we used a resolution of 60k for measurements of lower-abundance mass spectral fragments. This was not an issue in our Spring 2021 specimen, as the amino acid abundances were higher in each replicate measurement, and thus those measurements were performed with resolutions of 120k. Measurements of more abundant fragments were performed with Automatic Gain Control (AGC) targets of 2×10^5; less abundant fragments were performed with AGC targets of 5×10^4 (Supplemental Materials and Methods; Table S4.3).

Isotope ratio measurement

Carbon isotope ratios of mass spectral fragments of α-alanine (m/z = 140.0323 Da; referred to as the 140 fragment), β-alanine (m/z = 126.0166 Da and referred to as the 126 fragment), and aspartic acid (unsubstituted masses of 113.0239 Da, 156.0272 Da, 198.0378 Da, referred to as the 113, 156 and 198 fragments) from Murchison were measured relative to those of terrestrial standards (Supplemental Materials and Methods). With the Spring 2021 specimen, we measured the 113, 156, and 198 fragments of aspartic acid. With the Winter 2022 specimen, we measured the 156 fragment of aspartic acid, the 126 fragment of β-alanine and the 140 fragment of α-alanine.

Generally, to perform position-specific isotope ratio measurements on each sample, we required picomoles of the amino acid of interest for each replicate measurement of each fragment. To fully constrain each site of each amino acid, we needed to measure at least as many fragments are there were sites to constrain (to create a system of equations with an equal number of constraints and unknowns; Supplemental Materials and Methods). This was not always possible with all amino acids in both samples, as we were sample-limited, and the abundance of amino acids was variable between the samples (Table 4.1). With these limitations in mind, we prioritized measuring sites that most directly tested synthesis hypotheses (i.e., the amine sites).

In some cases, we compare across specimens and across studies to constrain additional molecular sites. We used the Winter 2022 sample to constrain the amine site of aspartic acid, but we report position-specific carbon isotope values for the first carboxyl, and the methylene+second carboxyl carbon sites of aspartic acid based on measurements of the Spring 2021 sample. The Winter 2022 sample was also used to measure the mass spectral fragments constraining the amine sites of α-alanine and β-alanine. Measurements of the same fragment of aspartic acid (156 Da) for both the Spring 2021 and Winter 2022 sample were used as a proxy for heterogeneity between the samples, and used to justify comparison between the $\delta^{13}C$ values of aspartic acid fragments from the two specimens. There was not enough Spring 2021 sample to measure fragments of α- or β-alanine. We only had enough sample to measure the amine site of α-alanine in the Winter 2022 site, so we use results from Chimiak, Elsila, et al., 2021 to constrain the methyl site of α-alanine. We contextualize our results with past CSIA of Murchison amino acids.

We compare sample fragment isotope ratios to those of amino acid standards with known molecular-average carbon isotopic composition. This comparison allows us to convert our measured isotope ratios into a standard reference frame analogous to the known international standard reference frame (i.e., VPDB for C) and report our data using standard carbon isotope delta notation, defined below (Equation 4.1). We note that to perform this conversion from our "house standard" reference frame to the international standard, we must make an assumption that there is no intramolecular carbon isotope heterogeneity in our amino acid standards; we elaborate further on this assumption in "Conversion to modeled VPDB (mPDB*) space and relevant assumptions."

We report fragment and position-specific isotope ratios in the common "delta" (δ) notation, calculated as the relative difference in isotope ratio between sample and standard, expressed in parts per thousand (per-mille; ‰):

$$\delta^A X = \left(\frac{{}^{A/a}R_{sample}}{{}^{A/a}R_{standard}} - 1 \right) \qquad (4.1)$$

where A and a are the cardinal masses of the rare and common isotopes of interest, respectively, X is the formula of the isotope-substituted element in question, and R is the observed ratio of interest.

Solving for position-specific isotope values

The molecular structures of targeted fragments were determined by examining their exact masses and comparing potential molecular structures for each mass with predictions from spectral interpretation software (Mass Frontier, Thermo Scientific). Contributions from particular sites within the parent molecule were confirmed by measuring aspartic acid and β-alanine labeled (10% ^{13}C label) at each atom position (Supplemental Materials and Methods). We derived the following equations following a matrix math similar to what was described in Wilkes et al., 2022 for the computation of the position-specific carbon isotope values of serine. We note here that the following equations assume that delta values can be used as proxies for isotopic fractional abundances and therefore mix conservatively; while not precisely correct, this assumption leads to errors of up to 4‰ for the measured delta values which are within the measured analytical uncertainties of this study.

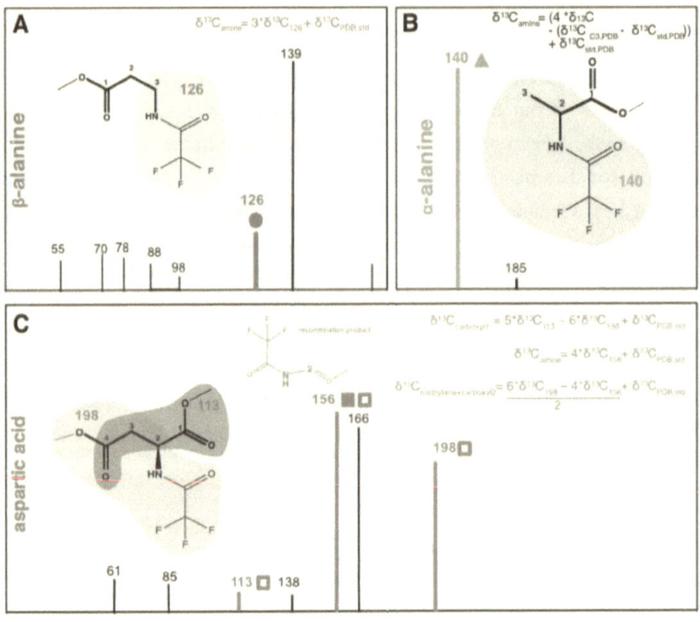

Figure 4.2: Amino acid mass spectral fragments.

Caption continued on next page.

Figure 4.2: Amino acid mass spectral fragments.

Derivatized amino acids form characteristic fragments when introduced via GC into the electron impact source. These fragments can be used to independently constrain the isotopic composition of distinct combinations of carbon sites and calculate position-specific isotope ratios. Here, we depict fragmentation spectra of (A) β-alanine, (B) α-alanine, and (C) aspartic acid. The derivatized amino acid structure is inset within each figure, where the bolded part of the structure highlights the structure of the underivatized amino acid. Shaded areas illustrate the molecular geometry of mass spectral fragments of the original derivatized molecule that we measured within this study, although some fragments we measured were recombination products (e.g., 156 fragment of aspartic acid) and thus are depicted alongside the original molecular structure. Teal circles represent β-alanine measurements, grey triangles represent α-alanine measurements and burnt orange squares represent aspartic acid measurements. For peak capture measurements, each colored fragment was isolated within and slowly purged from the reservoir into the mass spectrometer over minutes to tens of minutes (Figure 4.1). Fragment isotope ratios initially measured in comparison to the isotope ratio of house amino acid standards were converted into position-specific isotope ratios within a modeled PDB (mPDB*) reference frame using the equations presented within each of the panels (see "Isotope ratio measurement").

We identified one fragment (126) that allowed us to constrain the amine (C3) site of β-alanine (Figure 4.2A):

$$\delta^{13}C_{C3} = 3 \times \delta^{13}C_{126} + \delta^{13}C_{VPDB,standard} \qquad (4.2)$$

The other two sites of β-alanine can be constrained by measurement of the 139 Da fragment; we were not able to measure it due to prominent contaminants in the surrounding mass window, but include further description in the Supplementary Materials, as it would be a ripe target for future study.

The fragmentation spectrum of α-alanine was confirmed by Chimiak, Elsila, et al., 2021, who demonstrated that the amine and methyl (C2+C3) carbons can be constrained together by a measurement of the 140 Da fragment (Figure 4.2B). In that study, they reported $\delta^{13}C_{VPDB}$ value for the methyl (C3) carbon site of -36±20‰ for pristine Murchison. We had insufficient sample to complete a measurement of the other fragment necessary (m/z = 184.021) to constrain the methyl site alone. Thus, we use the value reported for the methyl (C3) carbon site of α-alanine by Chimiak, Elsila, et al., 2021 to compute a value for the amine (C2) site of alanine for our specimen as follows:

$$\delta^{13}C_{C2} = ((4 \times \delta^{13}C_{140}) - (\delta^{13}C_{amine} + \delta^{13}C_{VPDB,standard})) + \delta^{13}C_{VPDB,standard}$$

$$(4.3)$$

This choice is supported by similarities in calculated $\delta^{13}C_{VPDB}$ values of the amine site of alanine from our study and Chimiak, Elsila, et al., 2021's. Details regarding heterogeneity between Murchison specimens are described in the Discussion.

We identified three fragments of aspartic acid (113, 156, and 198) that permitted us to calculate position-specific values for the first carboxyl (C1) and amine (C2) sites of aspartic acid, and the methylene and second carboxyl (C3+C4) sites together (Figure 4.2C):

$$\delta^{13}C_{C1} = 5 \times \delta^{13}C_{113} - 6 \times \delta^{13}C_{198} + \delta^{13}C_{VPDB,standard} \qquad (4.4)$$

$$\delta^{13}C_{C2} = 4 \times \delta^{13}C_{156} + \delta^{13}C_{VPDB,standard} \qquad (4.5)$$

$$\delta^{13}C_{C3+C4} = 6 \times \delta^{13}C_{198} - 4 \times \delta^{13}C_{156} + \delta^{13}C_{VPDB,standard} \qquad (4.6)$$

Like β-alanine, the four sites could be fully constrained via measurement of the 138 Da fragment, but there were too many contaminants within the surrounding mass window to provide a confident measurement (Supplemental Materials and Methods).

The matrices and equations used for these position-specific computations for all three compounds are depicted in Figures 4.2 and S4.2. For each conversion, we assume that exogenous derivative atoms have identical isotopic compositions for both the sample and standard. This is a reasonable assumption because the sample and standard are derivatized at the same time under identical conditions and this derivative is known to fractionate C in a reproducible manner (Corr, Berstan, and Evershed, 2007; Silverman et al., 2021).

Errors for isotope ratio measurements are reported as standard errors across replicate acquisitions (σ_{SE}):

$$\sigma_{SE} = \frac{\sigma}{\sqrt{n}} \tag{4.7}$$

where σ is the standard deviation of the average isotope ratios for the set of replicate acquisitions, and n is the number of acquisitions. For position-specific isotope ratio measurements, σ_{SE} of the fragment delta values were propagated in quadrature following the fragment to position-specific matrix conversions (Supplemental Materials and Methods). Both σ and σ_{SE} are reported in ‰.

Conversion to modeled VPDB (mPDB) space and relevant assumptions*

It is important to contextualize our results within the already-established framework of the stable isotope geochemistry of carbon. As such, we converted position-specific $\delta^{13}C$ values measured within our house standard reference frame into the more widely used 'primary standard' isotope scales (e.g., VPDB; Brand et al., 2014), which we refer to as "modeled VPDB" (mPDB*), as demonstrated by Equation 4.6. To perform this conversion from our house standard reference frame to mPDB* reference frame, we added the measured compound-specific $\delta^{13}C_{VPDB}$ value of each standard as measured by the EA to our measured fragment delta values in the house standard reference frame. Similar to the assumption made for Equations 4.2-4.6 presented above, our conversion to mPDB* reference frame assumes conservative mixing of delta values.

Our mPBD* reference frame also incorporates an assumption that there is no intramolecular variation in carbon isotopic composition across the standard. This is not true for previously studied terrestrial amino acids; for instance, the first PSIA study by Abelson and Hoering, 1961 demonstrated via ninhydrin reaction of amino acids extracted from Chlorella pyrenoidosa that the $\delta^{13}C_{VPDB}$ values of the carboxyl sites of α-alanine and aspartic acid are higher — $\delta^{13}C_{VPDB}$ = 11.9‰ and 8.7‰, respectively — than the average of the other remaining sites — $\delta^{13}C_{VPDB}$ = -14.8‰ and -15.7‰, respectively (Abelson and Hoering, 1961). Intramolecular variations in carbon isotopic composition across commercially synthesized standards are even smaller (i.e., 10-15‰, constrained by via ninhydrin reaction; Chimiak, Elsila, et al., 2021), and within the errors that we report for each site. As such, we do not characterize these standard site-level differences here or incorporate them into our calculation of site-specific isotope ratios, but future studies could further improve the presentation of their isotope ratios on the international standard scale by adding additional measurements of the standard (based on EA-IRMS or nuclear magnetic resonance (NMR) constraints (e.g., Gilbert et al., 2009).

Results

Amino acid abundances

Abundances of L- and D-α-alanine from the Spring 2021 sample were 8.2±0.7 and 8.3±0.8 nmol/g, respectively. Abundances of L- and D-α-alanine from the Winter 2022 sample were 2.97±0.200 and 2.72±0.202 nmol/g, respectively. β-alanine abundances in the Spring 2021 and Winter 2022 samples were 16±1 and 9.128±0.280 nmol/g, respectively. Abundances of L- and D-aspartic acid from the Spring 2021 sample were 1.7±0.1nmol/g and 0.95±0.04, respectively. Abundances of L- and D-aspartic acid from the Winter 2022 sample were 0.76±0.017nmol/g and 0.48±0.004, respectively. Amino acid abundances for the Spring 2021 sample were reported originally in Glavin, Elsila, et al., 2020. Overall, abundances in the Winter 2022 sample were half or less of the abundance of the same amino acids measured in the Spring 2021 sample.

α-Alanine enantiomers in both the Spring 2021 and Winter 2022 samples have been observed in near-racemic proportions (L_{ee} = 0) within error (Table 4.1) indicating that these samples are minimally contaminated (Chimiak, Elsila, et al., 2021; Friedrich et al., 2019; Glavin, Elsila, et al., 2020). The L_{ee} of aspartic acid for these chips is consistent with a previous report from the same stone (Glavin, Elsila, et al., 2020), and has been proposed to be endogenous to the meteorite, though this interpretation could not be confirmed in that study by isotopic analysis of each enantiomer due to low abundances of aspartic acid within the measured sample. Nevertheless, the similarity in the $\delta^{13}C$ values of the 156 Da fragment ion of aspartic acid in both specimens examined in this study provides additional support that the amino acids in the two different specimens of Murchison are similar in origin and are endogenous to the sample; i.e., it is unlikely that the same amino acids from two distinct specimens of the same meteorite experienced equal amounts of contamination resulting in the same position-specific $\delta^{13}C$ values.

Sample	Amino Acid	Concent. (nmolg)	L$_{ee}$(%)	Molecular Avg $\delta^{13}C$	$\delta^{13}C_{amine}$ (‰, mPDB*, $\pm_s igma_{SE}$)	$\delta^{13}C$ acid 1 (‰, mPDB*, $\pm\sigma_{SE}$)	$\delta^{13}C$ avg methyl, methy-lene (‰, mPDB*, $\pm\sigma_{SE}$)	$\delta^{13}C$ acid 2 (‰, mPDB*, $\pm\sigma_{SE}$)
2021	L-α-alanine	8.2±0.7§	0§					–
2021	D-α-alanine	8.3±0.8§						–
2021	β-alanine	16±1§	–	–	–	–	–	–
2021	L-aspartic acid	1.7±0.1§	28±3§		-9±10	13±25	-32±17	
2021	D-aspartic acid	0.95±0.04§						
2022	L-α-alanine	2.97±0.200	4.4±7		+109±25			–
2022	D-α-alanine	2.72±0.202						
2022	β-alanine	9.128±0.280	–	+33±24				–
2022	L-aspartic acid	0.76±0.017	23±2	-14±5				
2022	D-aspartic acid	0.48±.004						
Chimiak et al. 2021	L-α-alanine	5.98±1.03	6.0±11	26±3	+142±20	-29±10	-36±20	–
Chimiak et al. 2021	D-α-alanine	5.30±0.88		25±3				

Table 4.1: Results.

Caption continued on next page.

Figure 4.2: Results.

Concentrations and isotope ratios of the amino acids in this study compared to Chimiak, Elsila, et al., 2021's results. $\delta^{13}C$ values are reported in modeled PDB reference frame (mPDB*; see methods for details) ± propagated 1σ standard errors. §Abundances for the Spring 2021 sample are based on data reported in Glavin, Elsila, et al., 2020. Prior molecular average $\delta^{13}C$ values of amino acids relevant to this study measured in hydrolyzed extracts of other specimens of Murchison are reported in Table S4.5.

Position-specific isotope values of Murchison amino acids

Fragment ion $\delta^{13}C$ values (Table S4.1) were converted into position-specific $\delta^{13}C$ values (Table S4.2), which were converted from the reference frame of a standard with known isotopic composition into our mPDB* reference frame (Figure 4.2 & 4.3, Table 4.1, Table S4.2). Our measurements constrained the amine carbon $\delta^{13}C$ values for α- and β-alanine as $\delta^{13}C_{mPDB*}$ = +109±21‰ and +33±24‰, respectively (Figure 4.3). These values are ^{13}C-enriched relative to any previously studied terrestrial amino acid. In contrast, the aspartic acid extract of the same specimen yielded an amine site $\delta^{13}C_{mPDB*}$ = -14±5‰ (Figure 4.3). The $\delta^{13}C$ value of the amine carbon of aspartic acid was equal within error to that of the terrestrial standard (Figure 4.3).

Our measurement of the $\delta^{13}C$ of α-alanine's amine site is equal within error to the value previously measured for a different sample of Murchison (+142±20‰; Figure 4.3; Chimiak, Elsila, et al., 2021), suggesting that we can directly compare data generated in these two studies, and specifically supports our use of the $\delta^{13}C$ value of the methyl site of α-alanine reported by Chimiak, Elsila, et al., 2021 ($\delta^{13}C_{VPDB}$= -36±20‰) to constrain the site-specific values of our measured α-alanine.

$\delta^{13}C$ values of the amine carbons of aspartic acid extracted from Spring 2021 and Winter 2022 specimens were within error of each other, which suggests some consistency between samples and supports our choice to use the measurements of the Spring 2021 sample to constrain aspartic acid's full position-specific carbon isotopic composition. Measurements of aspartic acid from the Spring 2021 specimen demonstrate that the average $\delta^{13}C$ of the methylene and second carboxyl carbon sites (not observed separately in this specimen) is $\delta^{13}C_{mPDB*}$ = -32±17 — lower than (but within 2 standard errors of) the terrestrial standard's — while the $\delta^{13}C$ of the first carboxyl carbon site is higher than the standard's ($\delta^{13}C_{mPDB*}$ = 13±25; Table 1; Figure 4.3).

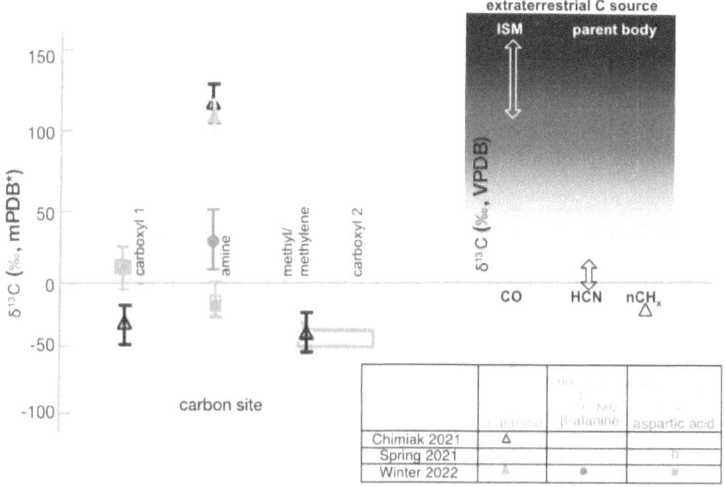

Figure 4.3: Position-specific carbon isotope values.

$\delta^{13}C$ values of carbon positions for α-alanine (this study = grey triangle; Chimiak, Elsila, et al., 2021 = open black triangle), β-alanine (teal circle) and aspartic acid (orange squares; open = Spring 2021, closed = Winter 2022) reported within our mPDB* reference frame. Error bars are $2\sigma_{SE}$. Key is inset within the table in the lower right. $\delta^{13}C$ values of different carbon sources within the ISM and on the meteorite parent body (Table S4.6) are plotted alongside for comparison, which have a range of isotopic compositions that could contribute to the amino acid's during synthesis.

Discussion

Our measured $\delta^{13}C$ value of α-alanine's amine site, averaged with the carboxyl and methyl sites reported in Chimiak, Elsila, et al., 2021 ($\delta^{13}C_{VPDB}$=-28.7‰ and -36.3‰, respectively) reproduces a molecular-average $\delta^{13}C$ value consistent within error with those measured in past CSIA of Murchison alanine (+25 to 52‰; Table S4.5; Chimiak, Elsila, et al., 2021; Elsila, Charnley, et al., 2012; Engel, Macko, and Silfer, 1990; Glavin, McLain, et al., 2020; Pizzarello, Huang, and Fuller, 2004). Our measured $\delta^{13}C$ value of β-alanine's amine site would be consistent with several measurements of the molecular-average $\delta^{13}C$ from prior CSIA of β-alanine ($\delta^{13}C_{VPDB}$= +5 to +10‰; Glavin, Elsila, et al., 2020; Pizzarello, Cooper, and Flynn, 2006; Pizzarello, Huang, and Fuller, 2004) if we assumed the isotopic composition of the combined methylene and carboxyl sites from β-alanine are comparable to that of the terrestrial standard ($\delta^{13}C_{VPDB} \sim$ -30‰), as is the case for the non-amine carbons in Chimiak, Elsila, et al., 2021's study of α-alanine (2021), as well as our methylene and second carboxylic acid carbons in the aspartic acid measured in the Spring 2021 sample. Our position-specific $\delta^{13}C$ values of aspartic acid, averaged across all four sites, are consistent within error with the molecular-average $\delta^{13}C$ of this same compound from Murchison ($\delta^{13}C_{VPDB} = $ +4‰), determined previously by CSIA (Table S5; Pizzarello, Krishnamurthy, et al., 1991). A prior CSIA by Pizzarello and others of the D- and L- enantiomers of aspartic acid reported $\delta^{13}C_{VPDB}$ values of -25.2 and -6.2‰, respectively, but they acknowledged lower $\delta^{13}C$ values for the L-enantiomers could be caused by terrestrial contamination (Pizzarello, Huang, and Fuller, 2004), which as we have already noted, we expect to be minimal or similar for our two measured Murchison specimens.

Our results suggest that α-alanine and β-alanine both formed from ^{13}C-enriched precursors, plausibly from the ISM, though with a significant difference in the final resulting $\delta^{13}C$ that potentially reflects distinct reaction pathways linking those precursors to final products. In contrast, aspartic acid appears to have formed entirely from precursors originating from relatively ^{13}C-depleted carbon sources (i.e., broadly resembling average meteoritic and terrestrial organic carbon); this implies that aspartic acid likely formed via a different reaction mechanism from both α and β-alanine, and possibly formed from precursors derived from the most common and abundant carbon components of the meteorite parent body. The similarity between the values of the methylene and second carboxylic acid carbons in aspartic acid and the methylene carbons in α-alanine further suggests the existence of shared carbon "pools" from which some carbons of different precursors derive. Indeed, prior work has suggested that carbon in meteoritic organic molecules are inherited from three potential sources: a ^{13}C-enriched CO pool from the ISM vs. relatively ^{13}C-depleted pools from HCN and nCH$_x$, possibly in the ISM and certainly on the meteorite parent body (Figure 4.3; Elsila, Charnley, et al., 2012).

Characterization of the isotopic compositions of carbon "pools" from which amino acids formed can be combined with known mechanisms of abiotic amino acid synthesis and isotope effects (IEs) imparted at each elementary reaction step to support or reject hypotheses regarding potential formation mechanisms. Reaction rates of elementary steps of irreversible reactions may differ between isotopic forms of reactants (kinetic isotope effects, or KIEs). Reversible exchange between reactants and products can result in equilibrium differences in isotopic content between the two (equilibrium isotope effects; EIEs). In both cases, IEs are quantified by the variable, ϵ, expressed as a difference in isotope ratio between reactant and product:

$$\epsilon^i_{j-k} = 1000 * (\alpha^i_{j-k} - 1) \qquad (4.8)$$

where α^i_{j-k} is the ratio of isotope ratios of the product and reactant, respectively, and the ϵ variable converts this α value into units of ‰. Negative ϵ^i_{j-k} values are "normal" isotope effects where the products are less heavy-isotope enriched than the reactants. These isotopic fractionations associated with reactions are specific to the reaction type and imprinted differently on different atomic sites of the product; thus, PSIA can help identify reaction types and therefore improve on constraints available from conventional isotope analyses.

Extraterrestrial Strecker synthesis of ^{13}C-enriched α-alanine

Both our study and the prior study by Chimiak, Elsila, et al., 2021 measured a $\delta^{13}C$ value of α-alanine's amine carbon that is highly ^{13}C-enriched compared to terrestrial organic carbon and molecular-average values of most C-rich compounds found in the carbonaceous chondrites. Chimiak, Elsila, et al., 2021 demonstrated that the PSIA of α-alanine from Murchison supports the hypothesis that it formed through Strecker-Cyanohydrin synthesis, whereby a distinctively ^{13}C-enriched amine carbon site was inherited from a ^{13}C-enriched carbonyl group of an aldehyde precursor (originally from interstellar CO), whereas the less ^{13}C-rich carboxyl and methyl sites of α-alanine were inherited from HCN and the CH_n moieties of aldehyde precursors, respectively.

We apply Chimiak, Elsila, et al., 2021's model (along with experimental data from Chimiak, Eiler, et al., 2022) to interpret our new measurement of the $\delta^{13}C$ value of the amine carbon of α-alanine; this approach lets us back-calculate the original $\delta^{13}C$ value of the precursor aldehyde molecule, applying the EIE (ϵ) for the addition of CN (average ϵ= -20‰; depicted in Figure 4.4A and originally reported in Chimiak, Eiler, et al., 2022; Chimiak, Elsila, et al., 2021) to solve for a $\delta^{13}C$ value of ~130‰ of the precursor carbon — the carbonyl site of reactant aldehyde. This result is consistent with prior studies that have measured and modeled higher $\delta^{13}C$ values of CO in the ISM ($\delta^{13}C$ = 110-160‰; Chimiak, Elsila, et al., 2021; Lyons, Gharib-Nezhad, and Ayres, 2018). We used this value as the starting point for ^{13}C-enriched amine sites for the rest of our modeled Strecker-related reactions (Figure

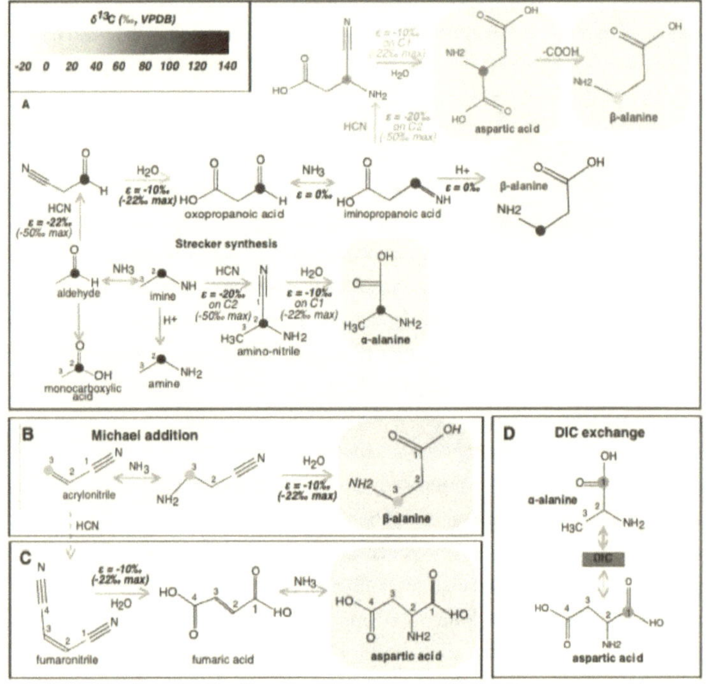

Figure 4.4: Potential extraterrestrial amino acid synthesis pathways, and ^{13}C enrichments of their amine sites.

Caption continued on next page.

4.4A). Hydrolysis also introduces an isotope effect on the reacting carbon (ϵ= -10‰ on C1 of α-alanine; Figure 4.4A), but it is within the range of our analytical error and the error reported by Chimiak, Elsila, et al., 2021 on the position-specific carbon isotope value of carboxyl carbon. The other reaction steps involved in Strecker synthesis of α-alanine do not have measurable IEs (Chimiak, Eiler, et al., 2022).

Figure 4.4: Potential extraterrestrial amino acid synthesis pathways, and ^{13}C enrichments of their amine sites.

(A) Extraterrestrial α-amino acids, including α-alanine, are thought to be synthesized via Strecker synthesis (grey box; Chimiak, Elsila, et al., 2021; Elsila, Charnley, et al., 2012), which inherits an ^{13}C-enriched carbon from interstellar medium CO in its amine carbon site (value derived from the $\delta^{13}C$ value of α-alanine's amine carbon). There are potential pathways to form β-alanine and aspartic acid through related reactions, but we rule out synthesis through these pathways from the same precursors of α-alanine and thus those pathways are shown in grey. (B) Michael addition of a nucleophile to an unsaturated carbonyl compound (acrylonitrile) is a common hypothesis for the formation of β-alanine (Burton et al., 2013; Elsila, Charnley, et al., 2012). For this pathway to be consistent with the $\delta^{13}C$ value we measured in β-alanine's amine site, the acrylonitrile precursor would need to be ^{13}C-enriched in its terminal carbon (Chimiak and Eiler, 2022). (C) Aspartic acid can also be synthesized via either amination of an unsaturated dicarboxylic acid or reductive amination of α-keto acid, likely via a fumaronitrile precursor. If all these carbons originated from the ^{13}C-depleted carbon pool on the meteorite parent body, the carbon isotope values of the amine site should not be noticeably higher than the values of other sites, as we observe in our results. It is possible that the fumaronitrile precursor may form from acrylonitrile (dotted arrow), but could not have the same enrichment as observed in β-alanine to be consistent with the value we measured for aspartic acid's amine site. (D) Carboxyl carbons could exchange with the carbonate in the host rock and become more ^{13}C-enriched.

A Strecker synthesis formation mechanism for α-alanine is supported by past CSIA (Aponte et al., 2017; Chimiak, Elsila, et al., 2021; Elsila, Charnley, et al., 2012); for instance, the observation that the $\delta^{13}C$ value of α-amino acids decreases with increasing number of carbon atoms (Sephton and Gilmour, 2000), which has been interpreted to reflect a dilution of a ^{13}C-enriched aldehyde precursor with added HCN/nCH$_x$-derived carbons (Chimiak, Elsila, et al., 2021; Chimiak and Eiler, 2022; Elsila, Charnley, et al., 2012). FTT synthesis is disfavored as a path to forming these compounds, as it would not create such distinctive differences in intramolecular isotopic composition, and is incapable of explaining the pattern of intermolecular differences in molecule average C isotope composition documented in many past studies (Chimiak, Elsila, et al., 2021). More work is necessary to understand isotopic fractionations related to amino acid formation via irradiation of methanol-containing ices, so our data and model arguments should not be construed as disproof of this alternative hypothesis.

Additional pathways required for synthesis of β-alanine and aspartic acid

Understanding the potential pathways to forming β-alanine and aspartic acid is more challenging, as there are diverse hypotheses as to how synthesis of these amino acids may occur and prior compound-specific carbon isotope data do not clearly discriminate among these possibilities. Three possible ways to synthesize β-alanine are established in the literature: reductive amination of an oxopropanoic acid intermediate that was itself formed via Strecker synthesis (Figure 4.4A); decarboxylation of aspartic acid (Figure 4.4A); and Michael addition of an unsaturated carbonyl (i.e., acrylonitrile) which is then aminated and hydrolyzed to form β-alanine (Figure 4.4B; Burton et al., 2013; Elsila, Charnley, et al., 2012; Peltzer et al., 1984).

The $\delta^{13}C_{mPDB*}$ value of β-alanine's amine carbon is enriched, but less enriched than that of α-alanine's amine carbon for the specimens measured in this study. We rule out synthesis exclusively by a Strecker synthesis mechanism (i.e., reductive amination of an oxopropanoic acid intermediate, where the amine carbon is inherited from interstellar ^{13}C-enriched CO; Figure 4.4A), as the same precursors that made the α-alanine measured here could not have synthesized a β-alanine that is consistent with our PSIA: β-alanine would have to be much more enriched.

Our findings are also inconsistent with β-alanine within Murchison forming exclusively via decarboxylation of aspartic acid, or at least based on the PSIA of aspartic acid measured in this study. Decarboxylation imparts an IE (ϵ= 2 to 40‰, depending on whether it is catalyzed or uncatalyzed; Lewis et al., 1993; O'Leary and Yapp, 1978), so if Murchison β-alanine formed this way from decarboxylation of Murchison aspartic acid, the $\delta^{13}C$ value of α-alanine's amine carbon should be similar to or lower than that of the aspartic acid it formed from, which is inconsistent with our results.

Thus, we propose that the β-alanine measured in this study was synthesized via Michael addition of a nucleophile to an unsaturated carbonyl compound (e.g., acrylonitrile; Figure 4.4B). The expected intramolecular isotopic composition of an acrylonitrile is debated in the literature, ranging from no expected intramolecular carbon heterogeneities (Elsila, Charnley, et al., 2012) to the formation of a nitrile where the terminal carbon could inherit a ^{13}C enrichment from interstellar CO (Figure 4.4B; Chimiak and Eiler, 2022). It is also possible that β-alanine's $\delta^{13}C$ value could represent a combination of synthesis processes, where β-alanine with a ^{13}C-enriched-amine site, produced through some combination of reactions in Figure

4.4A, is diluted by formation of β-alanine through processes resembling Michael addition (Figure 4.4B) or aspartic acid decarboxylation (Figure 4.4C) with smaller isotope effects. Additional position-specific isotope measurements of β- amino acids with longer carbon chains (e.g., β-amino n butyric acid) could help to further narrow down the most plausible synthesis mechanisms for this class of molecules.

Aspartic acid's lower $\delta^{13}C$ values at all measured carbon positions, particularly compared to the ^{13}C-rich amine carbons of α- and β-alanine, suggest that a distinct mechanism (or combination of mechanisms) must be responsible for its formation. Proposed mechanisms to form aspartic acid abiotically include: Strecker synthesis via an oxopropanoic acid intermediate (Figure 4.4A), or synthesis via either amination of an unsaturated dicarboxylic acid or reductive amination of α-keto acid, likely through a fumaronitrile precursor (Figure 4.4C; Burton et al., 2013; Pizzarello, Cooper, and Flynn, 2006).

It is possible that the aspartic acid measured in our specimen was formed via the Strecker pathway, but to be consistent with our measurements either the KIEs from two or more reaction steps would have to be maximally expressed, or the aspartic acid would need to be synthesized from aldehyde precursors with lower $\delta^{13}C$ values of the carbonyl site than the precursors responsible for synthesis of α-alanine (Simkus, Aponte, Hilts, et al., 2019). Thus, aspartic acid could not have formed by Strecker synthesis as part of the same reaction network with the same precursors that formed α- and β-alanine, but could have formed by Strecker synthesis from a separate set of precursors (perhaps in a different environment).

Alternatively, aspartic acid's ^{13}C-depleted amine, methylene and second carboxyl carbons are consistent with synthesis via a fumaronitrile precursor in the meteorite parent body, from the same HCN and nCH_x precursors that have been previously called on to explain synthesis of amino acids (Chimiak, Elsila, et al., 2021, and preceding sections of this paper). Precursors for such reactions that could have been present in the meteorite parent body include unsaturated dicarboxylic acids, which have been identified in Murchison (e.g., fumaric acid; Sephton, 2002). Prior studies have performed CSIA of dicarboxylic acids (Cronin et al., 1993; Pizzarello, Huang, Becker, et al., 2001), finding $\delta^{13}C$ values ranging from 28 to -6‰, similar to the range of measured $\delta^{13}C$ values for aspartic acid (Table S4.5). Average IEs associated with converting ^{13}C-depleted fumaronitrile precursors (Table S4.6) into aspartic acid would be -22‰ to each reacting carbon for the addition of water (Figure 4.4C), and thus result in aspartic acid without intramolecular ^{13}C enrichments (at the scale resolved in this study). Interestingly, synthesis via the fumaronitrile precursor has been posited as a plausible pathway to synthesize organics on early Earth via a cyanosulfidic protometabolism under subtly alkaline conditions (Patel et al., 2015).

Heterogeneity in the abundances and isotopic compositions of meteoritic organics

Organic chemical and stable isotopic heterogeneities among carbonaceous chondrites have long been documented. Amino acid molecular average $\delta^{13}C$ values vary depending on the specimen analyzed (Glavin, Callahan, et al., 2010). This isotopic variation may reflect spatial or temporal variations in isotopic compositions of precursors of organic compounds (e.g., $\delta^{13}C$ of interstellar CO varies from 110-160‰; $\delta^{13}C$ of aldehydes and ketones extracted from Murchison range from -10 to 66‰; Table S4.6; Chimiak, Elsila, et al., 2021; Lyons, Gharib-Nezhad, and Ayres, 2018; Simkus, Aponte, Hilts, et al., 2019, or a combination of synthesis processes, as discussed here.

Both our study and prior studies have highlighted variable amino acid abundances within distinct specimens of the same meteorite (Glavin, Callahan, et al., 2010 and data herein, Table 4.1). These variations in abundances could be driven by primary (i.e., formation) or secondary (i.e., redistribution and alteration) processes. Prior research has demonstrated that aqueous alteration can affect the abundance of amino acids, linking β-alanine abundance with degree of alteration, suggesting some amino acid formation mechanisms may be more productive with more abundant water (Botta, Martins, and Ehrenfreund, 2007). Likewise, aqueous alteration can alter the isotopic compositions of meteoritic amino acids, as isotopic variations can arise from differences in extents of isotopic exchange between the studied organic compounds and other parent body reservoirs (i.e., carbonate during aqueous alteration; Pietrucci et al., 2018). Different degrees of carbonate exchange in the carboxyl site of amino acids may be responsible for the heterogeneity in prior molecular average $\delta^{13}C$ measurements (Table S4.5) and to the values presented in our study. For instance, aspartic acid's first carboxyl site has a higher $\delta^{13}C$ value than the equivalent site in the terrestrial standards, which could be driven by exchange with dissolved inorganic carbon in parent body waters; carbonates in the aqueous alteration assemblage of the Murchison meteorite that have been demonstrated to be ^{13}C-enriched, reaching a maximum $\delta^{13}C_{VPDB}$ of 80‰ (Alexander, Kagi, and Larcher, 1982; Telus et al., 2019). Indeed, carboxyl-carbonate exchange was suggested by Chimiak, Elsila, et al., 2021 as one possible explanation of differences between the $\delta^{13}C$ value of the C1 carboxylic acid carbons of their method development and analytical samples. It is notable that the $\delta^{13}C$ values measured here and by Chimiak, Elsila, et al., 2021 for amino acid carboxyl sites have a range of values, but that none reach the maximum observed for carbonate grains in CCs, suggesting that even though exchange may be occurring, amino acid carboxyl sites do not reach full equilibrium with the dissolved inorganic carbon of parent body fluid.

Finally, sample preparation procedures may bias which amino acids are released from the meteorite sample via acid hydrolysis. Acid hydrolysis frees amino acids otherwise bound within larger polymers, and is used routinely in the preparation of meteoritic amino acids for analysis (Elsila, Charnley, et al., 2012; Simkus, Aponte, Elsila, et al., 2019). The starting materials for the hydrolyzed amino acids released from meteorites remain mostly unknown, and it is possible that the production of amino acids from unknown precursors could have an effect on the isotopic composition of the measured amino acids (Simkus, Aponte, Hilts, et al., 2019). However, one prior study demonstrated no appreciable difference in the compound-specific carbon isotopes of amino acids within two -CH_3 carbonaceous chondrites measured in the unhydrolyzed versus hydrolyzed extracts (Burton et al., 2013). It is also compelling that known isotope effects for each proposed reaction step in our posited mechanisms (Figure 4.4) can connect our measured position-specific isotope effects with previously measured isotope values of potential reactant molecules (Table S4.6).

Conclusions

Amino acids on meteorites are formed from precursors including interstellar components, through a heterogeneous and complex reaction network combining Strecker synthesis, Michael addition, and a fumaronitrile-based process that resembles posited mechanisms of prebiotic chemistry on Earth. Our observations are consistent with a scenario in which α-alanine formed entirely through Strecker synthesis, drawing on aldehyde precursors with a ^{13}C-enriched carbonyl site (presumably inherited from CO in the ISM), and ^{13}C-depleted non-carbonyl sites as well as ^{13}C-depleted HCN. In contrast, β-alanine could have formed via Michael addition drawing on both ^{13}C-enriched CO-derived and ^{13}C-depleted HCN precursors. Aspartic acid is most consistent with synthesis via a fumaronitrile precursor with no ^{13}C-enriched position-specific isotope values and thus likely derived from ^{13}C-depleted HCN. This network also calls on reaction mechanisms that require liquid water and neutral-to-mildly alkaline conditions, and thus could potentially occur (and compete with one another and other reactions) contemporaneously on the meteorite parent body.

This study has addressed one of several questions raised by the hypothesis in Chimiak, Elsila, et al., 2021 and Chimiak and Eiler, 2022, i.e., that a broad spectrum of soluble organics in the several of the carbonaceous chondrites share a synthetic history as components of a Strecker-based reaction network. Future studies should further test, modify and/or extend this hypothesis through additional PSIA of other meteoritic organics; we believe such work will make progress most quickly by focusing on branched amino acids, amines, and hydroxy acids, as the Chimiak, Elsila, et al., 2021 hypothesis explains their molecular average isotopic compositions through mechanisms that are predicted to leave specific, high-amplitude site-specific isotopic signatures. Understanding secondary alteration (pre- and post-fall) and its effects on contamination and alteration of intramolecular isotopic composition of meteoritic amino acids would also be invaluable to progressing this work. Development of methods for an improved understanding of extraterrestrial organic synthesis is more important now than ever; this study and others like it will be central in understanding the carbon isotopic composition, source and synthesis of any organics found through Mars Sample Return, NASA's OSIRIS Perseverance, OSIRIS-REx mission, JAXA's Hayabusa2 and MMX missions, and future sample return missions from organic-rich objects to elucidate the astrochemical and potential biochemical synthetic pathways beyond Earth.

Research data: Raw data from this study is available in an online repository, located at this DOI.

Author Contributions: Sarah S. Zeichner: Methodology, Validation, Formal analysis, Investigation, Data curation, Software, Writing, Visualization, Supervision, Project administration. Elle Chimiak: Methodology, Validation, Formal analysis, Investigation, Writing. Jamie Elsila: Methodology, Validation, Formal analysis, Writing (Review & Editing). Alex Sessions: Conceptualization, Methodology, Writing (Review & Editing). Jason Dworkin: Conceptualization, Formal analysis, Writing (Review & Editing). Jose Aponte: Conceptualization, Formal analysis, Writing (Review & Editing). John Eiler: Conceptualization, Methodology, Writing, Supervision, Resources, Funding acquisition.

References

Abelson, P.H. and T.C. Hoering (1961). "Carbon Isotope Fractionation of Amino Acids by Photosynthetic Organisms". In: *Proceedings of the National Academy of Sciences* 47, pp. 623–632.

Alexander, R., R.I. Kagi, and A.V. Larcher (1982). "Clay catalysis of aromatic hydrogen-exchange reactions". In: *Geochim Cosmochim Acta* 46, pp. 219–222.

Aponte, J.C. et al. (2017). "Pathways to meteoritic glycine and methylamine". In: *ACS Earth and Space Chemistry* 1, pp. 3–13. DOI: 10.1021/acsearthspacechem.6b00014.

Bernstein, M.P. et al. (2002). "Racemic amino acids from the ultraviolet photolysis of interstellar ice analogues". In: *Nature*, pp. 401–403.

Botta, O. and J.L. Bada (2002). "Extraterrestrial organic compounds in meteorites". In: *Surveys in Geophysics* 23, pp. 411–467. DOI: 10.1023/A:1020139302770.

Botta, O., Z. Martins, and P. Ehrenfreund (2007). "Amino acids in Antarctic CM1 meteorites and their relationship to other carbonaceous chondrites". In: *Meteoritics Planetary Science* 42. Retrieved from, pp. 81–92.

Brand, W.A. et al. (2014). "Assessment of international reference materials for isotope-ratio analysis (IUPAC technical report". In: *Pure and Applied Chemistry* 86, pp. 425–467.

Burton, A.S. et al. (2013). "Extraterrestrial amino acids identified in metal-rich CH and CB carbonaceous chondrites from Antarctica". In: *Meteoritics and Planetary Science* 48, pp. 390–402. DOI: 10.1111/maps.12063.

Cerling, Thure E. et al. (Sept. 1997). "Global vegetation change through the Miocene/Pliocene boundary". In: *Nature* 389 (6647), pp. 153–158. DOI: 10.1038/38229.

Charnley, S.B. et al. (2004). "Observational tests for grain chemistry: Posterior isotopic labelling". In: *Monthly Notices of the Royal Astronomical Society* 347, pp. 157–162. DOI: 10.1111/j.1365-2966.2004.07188.x.

Chimiak, L., J. Eiler, et al. (2022). "Isotope effects at the origin of life: Fingerprints of the Strecker synthesis". In: *Geochimica et Cosmochimica Acta* 321, pp. 78–98. DOI: 10.1016/j.gca.2022.01.015.

Chimiak, L., J.E. Elsila, et al. (2021). "Carbon isotope evidence for the substrates and mechanisms of prebiotic synthesis in the early solar system". In: *Geochimica et Cosmochimica Acta* 292, pp. 188–202. DOI: 10.1016/j.gca.2020.09.026.

Chimiak, Laura and J. Eiler (2022). "Prebiotic Synthesis on Meteorite Parent Bodies: Insights from Hydrogen and Carbon Isotope Models". In: *Nature* 389, pp. 153–158. DOI: 10.26434/chemrxiv-2022-3kzsz.

Cordiner, M.A. et al. (2019). "ALMA Autocorrelation Spectroscopy of Comets: The HCN/H^{13}CN Ratio in C/2012 S1 (ISON)". In: *Astrophys J Lett* 870, L26.

Corr, L.T., R. Berstan, and R.P. Evershed (2007). "Optimisation of derivatisation procedures for the determination of $\delta^{13}C$ values of amino acids by gas chromatography-combustion-isotope ratio mass spectrometry". In: *Rapid Communications in Mass Spectrometry* 21, pp. 3759–3771. DOI: 10.1002/rcm.3252.

Cronin, J.R. et al. (1993). "Molecular and isotopic analyses of the hydroxy acids, dicarboxylic acids, and hydroxydicarboxylic acids of the Murchison meteorite". In: *Geochimica et Cosmochimica Acta* 57, pp. 4745–4752. DOI: 10.1016/0016-7037(93)90197-5.

DeNiro, M.J. and S. Epstein (1977). "Mechanism of carbon isotope fractionation associated with lipid synthesis". In: *Science* 197, pp. 261–263. DOI: 10.1126/science.327543.

Eiler, J.M. et al. (2017). "Analysis of molecular isotopic structures at high precision and accuracy". In: *International Journal of Mass Spectrometry* 422, pp. 126–142.

Elsila, J.E., J.C. Aponte, et al. (2016). "Meteoritic amino acids: Diversity in compositions reflects parent body histories". In: *ACS Central Science* 2, pp. 370–379. DOI: 10.1021/acscentsci.6b00074.

Elsila, J.E., S.B. Charnley, et al. (2012). "Compound-specific carbon, nitrogen, and hydrogen isotopic ratios for amino acids in CM and CR chondrites and their use in evaluating potential formation pathways". In: *Meteoritics and Planetary Science* 47.9, pp. 1517–1536. DOI: 10.1111/j.1945-5100.2012.01415.x.

Engel, M.H., S.A. Macko, and J.A. Silfer (1990). "Carbon isotope composition of individual amino acids in the Murchison meteorite". In: *Nature* 348, pp. 47–49. DOI: 10.1038/348047a0.

Friedrich, J.M. et al. (2019). "Effect of polychromatic X-ray microtomography imaging on the amino acid content of the Murchison CM chondrite". In: *Meteorit Planet Sci* 54, pp. 220–228.

Gilbert, A. et al. (2009). "Accurate quantitative isotopic ^{13}C NMR spectroscopy for the determination of the intramolecular distribution of ^{13}C in glucose at natural abundance". In: *Analytical Chemistry* 81, pp. 8978–8985. DOI: 10.1021/ac901441g.

Glavin, D.P., C.M.O.D. Alexander, et al. (2018). "The origin and evolution of organic matter in carbonaceous chondrites and links to their parent bodies". In: *Primitive Meteorites and Asteroids. Elsevier*. Ed. by N. Abreu. DOI: 10.1016/B978-0-12-813325-5.00003-3.

Glavin, D.P. and J.P. Dworkin (2009). "Enrichment of the amino acid L-isovaline by aqueous alteration on CI and CM meteorite parent bodies". In: *Proceedings of the National Academy of Sciences* 106, pp. 5487–5492. DOI: 10.1073/pnas.0811618106.

Glavin, Daniel P., M.P. Callahan, et al. (2010). "The effects of parent body processes on amino acids in carbonaceous chondrites". In: *Meteoritics and Planetary Science* 45, pp. 1948–1972. DOI: `10.1111/j.1945-5100.2010.01132.x`.

Glavin, Daniel P., J.E. Elsila, et al. (2020). "Extraterrestrial amino acids and L-enantiomeric excesses in the CM2 carbonaceous chondrites Aguas Zarcas and Murchison". In: *Meteoritics Planetary Science* 26, pp. 1–26. DOI: `10.1111/maps.13451`.

Glavin, Daniel P., H. L. McLain, et al. (2020). "Abundant extraterrestrial amino acids in the primitive CM carbonaceous chondrite Asuka 12236". In: *Meteoritics and Planetary Science* 55 (9), pp. 1979–2006. DOI: `10.1111/maps.13560`.

Gordon, E.F. and D.C. Muddiman (2001). "Impact of ion cloud densities on the measurement of relative ion abundances in Fourier transform ion cyclotron resonance mass spectrometry: Experimental observations of coulombically induced cyclotron radius perturbations and ion cloud dephasing rates". In: *Journal of Mass Spectrometry* 36, pp. 195–203. DOI: `10.1002/jms.121`.

Hayatsu, R., M.H. Studier, and E. Anders (1971). "Origin of organic matter in early solar system?IV. Amino acids: Confirmation of catalytic synthesis by mass spectrometry". In: *Geochimica et Cosmochimica Acta* 35, pp. 939–951. DOI: `10.1016/0016-7037(71)90007-X`.

Hoegg, E.D. et al. (2018). "Concomitant ion effects on isotope ratio measurements with liquid sampling-atmospheric pressure glow discharge ion source Orbitrap mass spectrometry". In: *Journal of Analytical Atomic Spectrometry* 33, pp. 251–259. DOI: `10.1039/c7ja00308k`.

Hofmann, A.E. et al. (2020). "Using Orbitrap mass spectrometry to assess the isotopic compositions of individual compounds in mixtures". In: *Int J Mass Spectrom* 457, p. 116410.

Huber, C. and G. Wächtershäuser (2003). "Primordial reductive amination revisited". In: *Tetrahedron Letters* 44, pp. 1695–1697. DOI: `10.1016/S0040-4039(02)02863-0`.

Jørgensen, J. K. et al. (2018). "The ALMA-PILS survey: isotopic composition of oxygen-containing complex organic molecules toward IRAS 16293-2422B". In: *Astronomy Astrophysics* 620, A170. DOI: `10.1051/0004-6361/201731667`.

Koga, T. and H. Naraoka (2017). "A new family of extraterrestrial amino acids in the Murchison meteorite". In: *Scientific Reports* 7. DOI: `10.1038/s41598-017-00693-9`.

Kress, M.E. and A.G.G.M. Tielens (2001). "The role of Fischer-Tropsch catalysis in solar nebula chemistry". In: *Meteoritics and Planetary Science* 36, pp. 75–91. DOI: `10.1111/j.1945-5100.2001.tb01811.x`.

Lewis, C. et al. (1993). "Carbon Kinetic Isotope Effects on the Spontaneous and Antibody-Catalyzed Decarboxylation of 5-Nitro-3-carboxybenzisoxazole". In: *Journal of the American Chemical Society* 115.4, pp. 1410–1413. DOI: 10.1021/ja00057a025.

Lyons, J.R., E. Gharib-Nezhad, and T.R. Ayres (2018). "A light carbon isotope composition for the Sun". In: *Nature Communications* 9. DOI: 10.1038/s41467-018-03093-3.

Makarov, A. (2000). "Electrostatic Axially Harmonic Orbital Trapping: A High-Performance Technique of Mass Analysis". In: *Analytical Chemistry* 72, pp. 1156–1162.

Melzer, E. and M.H. O'Leary (1987). "Anapleurotic CO_2 Fixation by Phosphoenolpyruvate Carboxylase in C3 Plants". In: *Plant Physiology* 84, pp. 58–60. DOI: 10.1104/pp.84.1.58.

Miller, S.L. (1957). "The mechanism of synthesis of amino acids by electric discharges". In: *Biochimica et Biophysica Acta* 23, pp. 480–489. DOI: 10.1016/0006-3002(57)90366-9.

Monson, K.D. and J.M. Hayes (1980). "Biosynthetic control of the natural abundance of carbon 13 at specific positions within fatty acids in Escherichia coli. Evidence regarding the coupling of fatty acid and phospholipid synthesis". In: *Journal of Biological Chemistry* 255.23, pp. 11435–11441. DOI: 10.1016/s0021-9258(19)70310-x.

Muñoz-Caro, G.M. et al. (2002). "Amino acids from ultraviolet irradiation of interstellar ice analogues". In: *Nature* 416, pp. 403–406. DOI: 10.1038/416403a.

Neubauer, C. et al. (2018). "Scanning the isotopic structure of molecules by tandem mass spectrometry". In: *International Journal of Mass Spectrometry* 434, pp. 276–286.

O'Leary, M.H. and C.J. Yapp (1978). "Equilibrium carbon isotope effect on a decarboxylation reaction". In: *Biochemical and Biophysical Research Communications* 80.1, pp. 155–160.

Öberg, K.I. (2016). "Photochemistry and Astrochemistry: Photochemical Pathways to Interstellar Complex Organic Molecules". In: *Chem Rev* 116, pp. 9631–9663.

Patel, B.H. et al. (2015). "Precursors in a Cyanosulfidic Protometabolism". In: *Nature Chemistry* 7, pp. 301–307. DOI: 10.1038/nchem.2202.

Peltzer, E.T. et al. (1984). "The chemical conditions on the parent body of the Murchison meteorite: Some conclusions based on amino, hydroxy and dicarboxylic acids". In: *Advances in Space Research* 4, pp. 69–74. DOI: 10.1016/0273-1177(84)90546-5.

Pietrucci, F. et al. (2018). "Hydrothermal decomposition of amino acids and origins of prebiotic meteoritic organic compounds". In: *ACS Earth and Space Chemistry* 2, pp. 588–598. DOI: 10.1021/acsearthspacechem.8b00025.

Pizzarello, S., G.W. Cooper, and G.J. Flynn (2006). "The nature and distribution of the organic material in carbonaceous chondrites and interplanetary dust particles". In: *Meteorites and the Early Solar System* II, pp. 625–652. DOI: `10.2307/j.ctv1v7zdmm.36`.

Pizzarello, S., Y. Huang, L. Becker, et al. (2001). "The organic content of the Tagish Lake meteorite". In: *Science* 293, pp. 2236–2239.

Pizzarello, S., R.V. Krishnamurthy, et al. (1991). "Isotopic analyses of amino acids from the Murchison meteorite". In: *Geochimica et Cosmochimica Acta* 55.3, pp. 905–910. DOI: `10.1016/0016-7037(91)90350-E`.

Pizzarello, Sandra (2014). "The nitrogen isotopic composition of meteoritic HCN". In: *The Astrophysical Journal Letters* 796.2. DOI: `10.1088/2041-8205/796/2/L25`.

Pizzarello, Sandra, Y. Huang, and M. Fuller (2004). "The carbon isotopic distribution of Murchison amino acids". In: *Geochimica et Cosmochimica Acta* 68, pp. 4963–4969. DOI: `10.1016/j.gca.2004.05.024`.

Rasmussen, C. and D.W. Hoffman (2020). "Intramolecular distribution of $^{13}C/^{12}C$ isotopes in amino acids of diverse origins". In: *Amino Acids* 52, pp. 955–964. DOI: `10.1007/s00726-020-02863-y`.

Seargent, D.A. (1990). "The Murchison meteorite: Circumstances of its fall". In: *Meteoritics* 25, pp. 341–342.

Sephton, M.A. (2002). "Organic compounds in carbonaceous meteorites". In: *Natural Product Reports* 19.3, pp. 292–311. DOI: `10.1039/b103775g`.

Sephton, M.A. and I. Gilmour (2000). "Aromatic Moieties in Meteorites: Relics of Interstellar Grain Processes?" In: *Astrophys J* 540, pp. 588–591.

Silverman, S.N. et al. (2021). "Practical considerations for amino acid isotope analysis". In: *Organic Geochemistry* 164, p. 104345. DOI: `10.1016/j.orggeochem.2021.104345`.

Simkus, D.N., J.C. Aponte, J.E. Elsila, et al. (2019). "Methodologies for analyzing soluble organic compounds in extraterrestrial samples: Amino Acids, Amines, Monocarboxylic Acids, Aldehydes, and Ketones". In: *Life* 9. DOI: `10.3390/life9020047`.

Simkus, D.N., J.C. Aponte, R.W. Hilts, et al. (2019). "Compound-specific carbon isotope compositions of aldehydes and ketones in the Murchison meteorite". In: *Meteoritics and Planetary Science* 54 (1), pp. 142–156. DOI: `10.1111/maps.13202`.

Su, X., W. Lu, and J.D. Rabinowitz (2017). "Metabolite spectral accuracy on Orbitraps". In: *Analytical Chemistry* 89, pp. 5940–5948. DOI: `10.1021/acs.analchem.7b00396`.

Telus, M. et al. (2019). "Calcite and dolomite formation in the CM parent body: Insight from in situ C and O isotope analyses". In: *Geochimica et Cosmochimica Acta* 260, pp. 275–291. DOI: 10.1016/j.gca.2019.06.012.

Uechi, G.T. and R.C. Dunbar (1992). "Space charge effects on relative peak heights in Fourier transform-ion cyclotron resonance spectra". In: *Journal of the American Society of Mass Spectrometry* 3, pp. 734–741.

Wilkes, E.B. et al. (2022). "Position specific carbon isotope analysis of serine by gas chromatography/Orbitrap mass spectrometry, and an application to plant metabolism". In: *Rapid Communications in Mass Spectrometry* 36. DOI: 10.1002/rcm.9347.

Zeichner, S.S. (2021). szeichner/DirectElution: DirectElutionv1.0.0. DOI: 10.22002/D1.2182.

Zeichner, S.S. et al. (2022). "Methods and limitations of stable isotope measurements via direct elution of chromatographic peaks using gas chromotography-Orbitrap mass spectrometry". In: *Int J Mass Spectrom* 477, p. 116848.

Amino acid standards

We prepared pure amino acid standards of L+D-aspartic acid (Sigma Lots 060M0092V and MKCC0521, respectively), β-alanine (Sigma Lot BCBR6059V), and L+D α-alanine (Sigma Lots BCBF7865V, SLBN6572V, respectively). The molecular-average carbon isotopic composition of the amino acid standards were characterized by Elemental Analyzer-Isotope Ratio Mass Spectrometry (EA-IRMS) at Caltech. The mean $\delta^{13}C_{VPDB}$ values ($\pm$ 1 σ_{SE}) of the aspartic acid, β-alanine, and α-alanine standards were -22.14 $\pm$ 0.07‰ (n = 3), -27.09 $\pm$ 0.40‰ (n = 3) and -19.60 $\pm$ 0.24‰ (n=3), respectively (Table S4.1). In addition, we prepared aspartic acid with 10% labels at each carbon site to verify which fragment ions inherit which carbon sites (gravimetrically mixed from natural abundance amino acids with >99% ^{13}C-labeled amino acids purchased from Sigma). We used a similar method to perform site assignments for β-alanine mass spectral fragments, however we were unable to acquire labeled compounds for carbon sites 2 and 3 separately, and thus had to characterize them together. Chimiak, Elsila, et al., 2021 followed a similar method to characterize the fragmentation spectrum and site-assignment of α-alanine. Finally, we prepared a "meteorite standard" with 34 amino acids scaled to relative abundances of reported in Glavin, Elsila, et al., 2020 for optimizing chromatography and GC oven programs. All standards and samples were derivatized as described below.

Overview of GC-Orbitrap mass spectrometry

Analyses were performed on a Q-Exactive Orbitrap mass spectrometer with samples introduced via a Trace 1310 gas chromatograph equipped with a TG-5SIL MS column (30 m, 0.25 mm ID, 0.25 μm thickness) with a 1 m fused silica capillary pre-column (inner diameter of 220 μm/outer diameter of 363 μm, VSD tubing) (Figure 4.1). To perform measurements of the Winter 2022 sample, we installed two TG-5SIL MS columns connected in series with a Valco union to improve separation between our compounds of interest and near co-eluting peaks. Samples and standards were analyzed under closely matched experimental conditions.

Compounds were injected via split-splitless injector operating in splitless mode, with He as a carrier gas with a flow rate of 1.2 mL/min. The GC oven temperature program was optimized to separate the analytes of interest from the rest of the organic compounds in the extract. The mass windows for each of the mass spectral fragments measured for PSIA are presented in Fig. S4.1. The GC has been modified through the addition of two, four-way valves and a peak-broadening reservoir, as described in Eiler et al., 2017 (Figure 4.1). For this study, we measured compounds in two different valve configurations: (1) "direct elution" of the GC effluent into the mass spectrometer (Zeichner et al., 2022) and (2) "peak capture" of the compound of interest within a peak-broadening reservoir (Eiler et al., 2017). Direct elution measurements were used in several ways. First, we used direct elution measurements to confirm the time of elution for each compound of interest. Direct elution measurements were also used to quantify the abundance of amino acids in both of the blanks compared to the intensity of the amino acid peaks in the sample (Table S4.4). Finally, direct elution measurements were critically important in detecting any potential contaminants, or near co-eluting compounds, that ionized to produce fragments within the mass window of interest. When possible, chromatography was optimized to minimize the presence of the contaminant ions that would compete with the mass fragment we were measuring (i.e., the addition of the second column for the measurement of the Winter 2022 sample). In some cases, even with an improvement in chromatography, we could not get a clean enough mass window to perform adequate isotope ratio measurements of the fragment of interest (e.g., 198 fragment in the Winter 2022 sample). Future studies wanting to apply this technique to perform PSIA of analytes within complex sample mixtures may want to explore preparatory offline clean-up steps for their samples prior to direct elution. The potential effects of these extraneous fragment ions, and potential strategies to mitigate them are further detailed below in our discussion of "space charge effects" under "Position-specific isotope ratio analysis."

To perform fragment isotope ratio measurements for PSIA, we used the "peak capture" method. Peak capture measurements enable the isolation of a single analyte from a complex mixture of compounds by turning valves to 'trap' the chromatographic peak of interest in a reservoir as it elutes off the GC column. Here, we used a silicon coated stainless steel reservoir (10 cc; SilcoTek, Bellefonte, PA). Following peak capture, the compound was slowly purged from the reservoir with helium over tens of minutes, permitting many different Orbitrap scans of a given sample, and thus improving the precision and accuracy of the measurement. With the peak capture method, each of our fragment ions eluted for minutes to tens of minutes, with a usable stable measurement time (i.e., the time frame where scans observed enough signal of the analyte of interest to use for isotope ratio measurement) of 3 to 10 minutes depending on the total concentration of the analyte introduced into the reservoir.

GC effluent was transferred directly into the ion source via a heated transfer line (260°C) where analytes were ionized via electron impact (EI; Thermo Scientific Extractabrite, 70eV; Fig. 4.1). Each compound generated a characteristic combination of fragment ions upon ionization (Figure 4.2), which were extracted from the source, subjected to collisional cooling during transfer through the bent flatpole, underwent mass-window selection by the Advanced Quadrupole SelectorTM (AQS), and were then passed through the automatic gain control (AGC) gate prior to storage in the C-trap — a potential-energy well generated by radio frequency and direct current potentials (Figure 4.1). Ions accumulated in the C-trap until the total charge reached a user-defined threshold (the "AGC target"), then were introduced into the Orbitrap mass analyzer as a discrete packet. Within the mass analyzer, ions orbited between a central spindle-shaped electrode and two enclosing outer bell-shaped electrodes, moving harmonically at frequencies proportional to m/z (Makarov, 2000). The raw data product of this oscillation—the "transient"—was converted via fast Fourier transform into a data product that can be processed for isotope ratio analysis (see Data processing).

We refer to each injection of a sample or standard as an "acquisition;" each acquisition is comprised of "scans," with each scan comprising ion intensities and m/z ratios averaged by the Orbitrap over a short time interval (typically 100-300 ms, where scan length is dependent on the user-defined nominal resolution setting). We refer to each set of replicate acquisitions of a fragment for both a sample and standard as an "experiment." It is critical that any application of the Orbitrap to isotope ratio analysis performs measurement under "AGC control," i.e., under conditions where the AGC target limits the number of ions admitted into the Orbitrap in each scan and is typically defined by low variation in the total ion current times the injection time (TIC×IT) over the course of a single acquisition. In our measurements, this was a constraint due to the low abundance of analyte; it was evident when the measurement when the instrument was not performing measurements under AGC control when the IT was at its maximum (3000 ms).

Position-specific isotope ratio analysis

Based on the results of labeling studies and direct elution measurements to evaluate near-coeluting peaks and the presence of contaminant ions in the system, we identified 3 target fragments for aspartic acid that independently constrain all four carbon sites (with unsubstituted masses of 113.0239 Da, 156.0272 Da, 198.0378 Da, referred to henceforth as the 113, 156 and 198 fragments) and 1 target fragment for β-alanine that constrains the amine site directly (m/z = 126.0166 Da and referred to henceforth as the 126 fragment) (Figure 4.2C&A). The fragment to constrain the amine site (140.0323 Da; referred to as the 140 fragment) of α-alanine was determined in a previous study (Figure 4.2B; Chimiak, Elsila, et al., 2021).

With the Spring 2021 specimen, we measured the 113, 156 and 198 fragments of aspartic acid, and the 126 fragment of β-alanine. With the Winter 2022 specimen, we measured the 156 fragment of aspartic acid, the 126 fragment of β-alanine and the 140 fragment of α-alanine. We attempted a measurement of the 198 fragment using the Winter 2022 specimen, but noticed too much evidence of contamination from the sample matrix for the measurement to be useful. The abundance of aspartic acid in the Winter 2022 specimen was too low abundance to attain a useful measurement of the 113 fragment. Each of these fragment measurements can be translated into position-specific isotope values.

To measure each fragment, we first performed a direct elution measurement of a standard to confirm the timing of the compound of interest. We identified compound chromatographic peaks by when the characteristic fragment ions appeared and disappeared above baseline in the mass spectra. We chose a time window around the elution of each chromatographic peak as the part of the GC effluent to capture in the reservoir and verified this timing by "reverse peak capture" of pure amino acid standards where the effluent was routed to vent for the time frame of interest and thus verified that the compound fragment ions disappeared entirely from the mass spectrum. Generally, we turned valves to begin peak capture 15-20 seconds prior to the arrival of the peak of interest, and waited until the intensity of the peak tail returned to the baseline NL score prior to peak elution to end peak capture.

Once we verified the time window, we performed a set of peak capture measurements, alternating between measurements of sample and standard whose relative concentration were adjusted so that they had approximately the same NL score in the reservoir. Target sample sizes corresponded to ~10-50 picomoles of derivatized amino acid to achieve maximum absolute peak intensities (which are referred to within the Orbitrap software and RAW data files as "NL scores") that were >1 order of magnitude above the reservoir baseline NL score for the fragment of interest

(~1 – 5×10^5 for the monoisotopic peak, and 5×10^3 – 1×10^4 for the background). Measurements of standards at different concentrations allowed us to anticipate the approximate achievable precision of a given amount of analyte, as we performed all our measurements under stable instrument conditions where achievable precision followed shot noise limits (see Data processing).

We selected AQS windows, resolutions, and AGC targets to achieve the maximum sensitivity for the fragment ion of interest (i.e., for the given experimental conditions, the majority of the ions being measured are the compound of interest; Eiler et al., 2017). However, introducing too many ions into the C-trap or the Orbitrap at once can lead to "space charge effects," a phenomenon by which ions affect the movement and trajectory of one another, disturbing them from the persistent, harmonic orbits required for Fourier-transform mass spectrometry (Eiler et al., 2017; Gordon and Muddiman, 2001; Hoegg et al., 2018; Hofmann et al., 2020; Neubauer et al., 2018; Su, Lu, and Rabinowitz, 2017; Uechi and Dunbar, 1992). Space charge effects can be mitigated for some cases by reducing extraneous ions in the mass window of interest so that there are fewer ions interacting with and capable of suppressing the isotope ratio of the fragments of interest (i.e., narrow the AQS window, lower the AGC target), or by reducing the length of time that the ions spend in the Orbitrap and thus the length of time they have to interact and that their harmonic orbits can inconsistently decay (i.e., lower resolution).

We performed all measurements with the narrowest AQS window that we could without introducing drops in transmission for fragments near the edge of the mass window (Eiler et al., 2017). Narrower mass windows were particularly helpful for this sample due to the complexity of the sample matrix and the large number of extraneous ions that often fell within the mass window that were not from our compound of interest. For the Spring 2021 specimen, we used AQS windows of 7 Da centered around the masses of the monoisotopic and substituted fragments of interest; for the Winter 2022 specimen we used AQS windows of 6.5 Da.

The resolution required to distinguish between the masses of closely adjacent near-isobars expected to be present in the mass spectrum can be calculated as:

$$resolution = \frac{m}{\Delta m} \tag{S4.1}$$

where m is the mass of an ion of interest and Δm is difference in mass necessary to separate that ion peak from an adjacent near-isobar. This required resolution is similar to the "nominal" resolution reported for the Orbitrap, which is calculated for a 200 Da fragment ion as m/dm, where m is mass and dm is the full peak width at half maximum intensity (FWHM). In general, sensitivity improves with lower resolution, which has also been shown to mitigate the effects of contaminants on the accuracy of isotope ratio measurements (Hofmann et al., 2020). Indeed, we did observe in preliminary experiments that with the presence of many extraneous ions within the mass window of interest, the isotope ratio of the sample relative to the standard was suppressed when we used a resolution of 120,000 instead of 60,000. All Spring 2021 sample measurements were performed with resolution of 120,000, which were measured at higher concentrations and therefore had fewer issues with contaminant ions of a similar concentration to the fragment ions of interest. All of the results we report from Winter 2022 analyses were performed using a resolution of 60,000.

The choice of AGC target was directly related to the relative abundance of the fragment of interest. Measurements of larger fragments with cleaner mass windows were performed with AGC targets of 2×10^5; smaller fragments (i.e., 113 fragment of aspartic acid) that may be more susceptible to space charge effects were measured using an AGC target of 5×10^4. All acquisitions for a single experiment were performed using the same parameters. A full set of experimental conditions used for each experiment is included in Table S4.3.

Data processing

Data files from each acquisition are converted to selected-mass chromatograms (intensity versus time for a selected m/z) which can then be integrated to yield isotope ratios. We used FT Statistic, a computer program written by ThermoFisher^{TM}, to extract data from RAW files created by the proprietary Orbitrap control software. From the FT Statistic-processed files, we extracted ion intensities (NL scores), peak noise, total ion current (TIC), injection time (IT), and other acquisition statistics, which we analyzed via in-house code written in Python (version 3.7.6). Definitions of each of these parameters can be found in prior Orbitrap studies (e.g., Eiler et al., 2017). The data analysis process is described in the subsequent paragraphs, and the processing code can be found on the Caltech data repository (Zeichner, 2021). Raw data for samples and standards were processed identically to enable comparisons between measurements, and are available in an online repository.

We went through the raw files for each acquisition of an experiment to identify a consistent time frame of integration where the Orbitrap mass analyzer was operating under AGC control. We identified this time frame based on when the IT time was < 3000 ms, or the maximum IT time. As a secondary check on AGC control, we confirmed that the TIC*IT variation was < 20% for all the scans used in our isotope ratio analysis. In many cases, different acquisitions of a given experiment stayed under AGC control for different lengths of time, likely due to minor inconsistencies in the overall abundance in analyte concentration from acquisition to acquisition. We used the most conservative (i.e., shortest) time frame for all acquisitions within an experiment. Prior studies that have used the Orbitrap to perform isotope ratio measurements have used a NL score-based culling threshold of 10% to decide which scans to use in isotope ratio measurements (Wilkes et al., 2022; Zeichner et al., 2022). However, we noticed that this threshold was not as applicable to the study of such small sample sizes; the time frame where our measurement of fragment ions by reservoir elution left AGC control ranged in what the intensity of the unsubstituted fragment was relative to the total ion current (NL score/TIC; e.g., 3-40%). For cases where the NL score / TIC < 5% when the measurement was no longer under AGC control, we used a 5% baseline as a cutoff.

We explored the use of baseline corrections for our measurements. In all cases, the baseline for the Murchison sample was higher than the standard, and so we expect that baseline subtraction would increase the isotope ratio of our samples relative to our standards. However, to perform adequate baseline corrections, it is critical to measure the background under AGC control. The background was only observed under AGC control for one experiment—the measurement of the 126 fragment—and otherwise the background prior to the elution of the analyte peak out of the reservoir was observed under conditions of maximum IT. We perform a background correction for the 126 fragment, where we subtract an average of the NL score for unsubstituted and the substituted fragment ion masses during the background scans from each of the scans selected within our chosen time frame for data analysis, prior to conversion to ion counts, described below. Note that the incorporation of this background correction for the 126 fragment measurement did increase the σ_{SE} due to having fewer counts to quantify and use in the computation of the isotope ratio of the fragment. For all other fragments, we report the relative height of the NL score of the background to that of the eluting peak (Table S4.3), but do not perform any quantitative baseline correction.

Raw data files report signal intensities, which must be converted into the number of ions ('ion counts') to compute isotope ratios. To calculate isotope ratios for each chromatographic peak, we next converted NL scores for each remaining scan into ion counts (Eiler et al., 2017):

$$N_{io} = \frac{S}{N} \times C_N/z \times \sqrt{\frac{R_N}{R}} \times \sqrt{\mu} \qquad \text{(S4.2)}$$

where N_{io} is the number of observed ions, S is the reported signal intensity (NL score) for the molecular or fragment ion in question, N is the noise associated with that signal, R is the formal resolution setting (defined for m/z 200) used, R_N is a reference formal mass resolution at which eqn.2 was established, C_N is the number of charges corresponding to the noise at reference resolution R_N (4.4; a constant established by prior experiments; see Eiler et al., 2017), z is the charge per ion for the fragment of interest, and μ is the number of microscans (i.e., sequential Orbitrap ion packet injections that were averaged, analyzed by fast-Fourier transform, and reported as a single scan).

To compute isotope ratios, we take scans where the monoisotopic peak intensity (NL score) is under AGC control, as described above. The unsubstituted and monoisotopic ions in the remaining scans were respectively summed, and then divided to calculate the isotope ratio of interest. We report both fragment and position-specific isotope ratios in the common "delta" (δ) notation (Equation 4.1). We explicitly state the reference frame for each reported value and use subscripts on delta values to denote the compound or carbon position for which the isotope value is being reported. With our delta values, we report the standard error for isotope ratio measurements across replicate acquisitions (σ_{SE}; Equation 4.7). Both σ and σ_{SE} are reported in per-mille (‰). To refer to σ_{SE} of specific fragments, "f", we refer to them as σ_{SE-f}.

We compare our achieved σ_{SE} with the theoretical "shot noise limit" of each acquisition based on the number of observed counts :

$$\frac{\sigma_{SNL}}{R} = \sqrt{\frac{1}{\sum C_{io}} + \frac{1}{\sum c_{io}}} \qquad \text{(S4.3)}$$

where $\frac{\sigma_{SNL}}{R}$ is the shot noise limit on a single acquisition, which is calculated based on $\sum C_{io}$ and $\sum c_{io}$ — the sums of all the counts for the substituted and unsubstituted molecular or fragment ions of interest, respectively. The shot noise limits represent the best-case uncertainty on a measurement of the N/n ratio and serves as a useful point of comparison with acquisition errors and experimental reproducibility, σ and σ_{RSE}, as defined above.

Fragments constraining other molecular positions

We attempted to measure two additional fragments: a 138.01607 Da fragment of aspartic acid that would allow us independently constrain the isotope values of the methylene and second carboxyl sites, and a 70.082874 and 139.0238 Da fragment of β-alanine that would allow us to constrain the isotope value of the methylene and carboxyl (C1+C2) carbon sites together. However, these masses fell within mass windows that often had fragments from other co-eluting and near-co-eluting compounds. It was impossible to get an adequately clean mass window for suitable isotope ratio measurements of these fragments (i.e., mass windows where there were not prominent (>20% of the NL score of the base peak) contaminant fragment ions), but future studies may be able to achieve this measurement with improved chromatography or offline preparatory separation steps.

The matrices used and equations used for these position-specific computations for all three compounds are depicted in Fig. S4.2. Fig. S4.2 also includes additional equations that could be used to constrain all of the carbon sites in each molecule, given additional measurements of 138 Da fragment of aspartic acid, 70 and 139 Da fragments of β-alanine and the 184 fragment of α-alanine.

Error propagation

Because each position-specific isotope value is measured indirectly, we must propagate the errors from fragment isotope ratio measurements, which we did for each fragment of aspartic acid as follows:

$$\sigma_{SE-C1} = \sqrt{(5\sigma_{SE-113})^2 + 6\sigma_{SE-198})^2} \qquad (S4.4)$$

$$\sigma_{SE-C2} = 4\sigma_{SE-113} \qquad (S4.5)$$

$$\sigma_{SE-C3+C4} = \sqrt{(6\sigma_{SE-198})^2 + 4\sigma_{SE-156})^2} \qquad (S4.6)$$

We note that errors on our position-specific isotope ratios are correlated due to the use of each fragment measurement in multiple calculation. Errors on the position-specific carbon isotope value of the amine (C3) carbon of β-alanine was calculated as follows:

$$\sigma_{SE-C3} = 3\sigma_{SE-126} \tag{S4.7}$$

To compute the error on the position-specific carbon isotope value of the amine carbon (C2) of α-alanine, we added errors reported by Chimiak, Elsila, et al., 2021 with propagated errors from our 140 fragment measurements of the sample and standards added in quadrature as follows:

$$\sigma_{SE-C2} = \sigma_{SE-C3,Chimiak} + 4\sqrt{(\sigma_{SE-140-standard})^2 + 4\sigma_{SE-140-sample})^2} \tag{S4.8}$$

Supplemental Tables and Figures

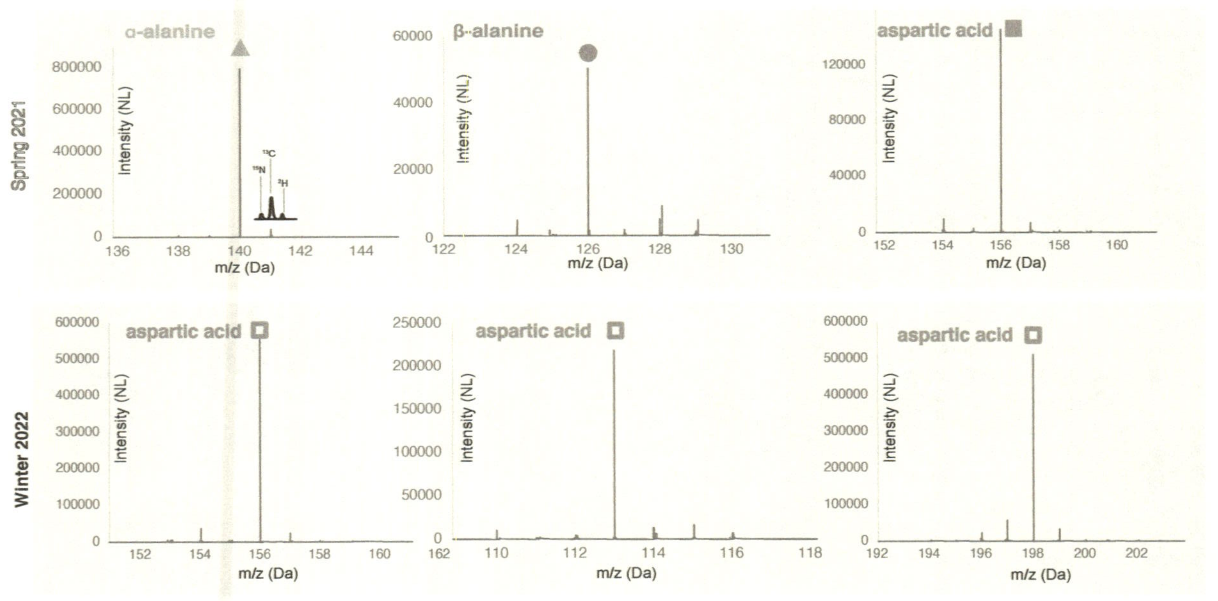

Fig. S4.1: Mass windows of measured mass spectral fragments.

Peak capture experiments measured narrowed mass windows centered around the unsubstituted mass fragment of interest.

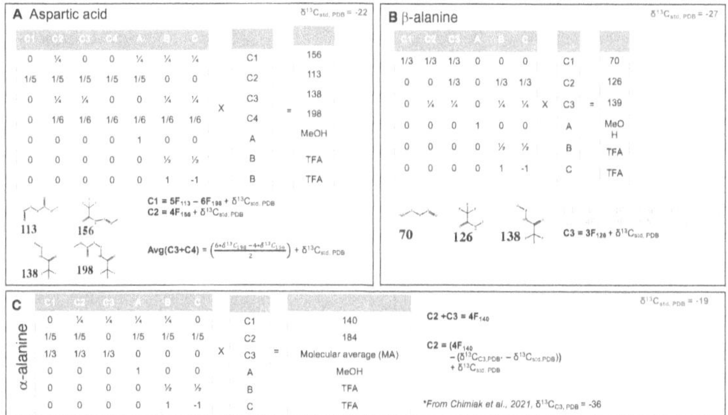

Fig. S4.2: Matrix math and equations to compute position-specific $\delta^{13}C_{VPDB}$ values.

(A) Aspartic acid matrix math and equations. (B) β-alanine matrix math and equations. (C) α-alanine matrix math and equations. All equations are written to compute full conversion of position-specific isotope values within mPDB* space (computed by adding the $\delta^{13}C_{VPDB}$ value of the house standard, included in the upper right corner of each panel in blue). Properties and equations used in this study are bolded and in black; other equations that could be used to fully constrain all sites of each molecule are included for completeness but in grey.

Amino acid	standard $\delta^{13}C \pm \sigma_{SE}$ (‰, VPDB)	Fragment	Specimen	Carbon sites	R_{sample}/R_{std} (σ_{SE})	Murchison $\delta^{13}C \pm \sigma_{SE}$ (‰, house standard)
Aspartic acid	-22.14±0.07	113	Spring 21	A,B,C,D	1.0055 (0.004)	5.5±4.1
		156	Spring 21	B	1.0033 (0.002)	3.3±2.4
		156	Winter 22	B	1.0020 (0.001)	2.0±1.4
		198	Spring 21	C,D	0.9987 (0.002)	-1.3±2.3
β-Alanine	-27.09±0.04	126	Winter 22	B	1.020 (0.008)	20.1±7.9
α-Alanine	-19.60±0.24	140	Winter 22	A,B	1.028 (0.0003)	27.7±1.2

Table S4.1: Mass spectral fragment isotope ratios and delta values.

Mass spectral fragment isotope ratios for aspartic acid, β-alanine and α-alanine from Murchison, reported as both $R_{sample}/R_{standard}$ and $\delta^{13}C_{housestandard}$. Values are reported along with molecular average $\delta^{13}C_{VPDB}$ of amino acid standards, measured by EA. Reported errors are $1\sigma_{SE}$.

Carbon letter	Specimen	Carboxyl 1			Amine			Avg. of + methylene carboxyl 2		
		$\delta^{13}C$ (‰, vs std)	$\delta^{13}C$ (‰, mPDB*)	σ_{SE}	$\delta^{13}C$ (‰, vs std)	$\delta^{13}C$ (‰, mPDB*)	σ_{SE}	$\delta^{13}C$ (‰, vs std)	$\delta^{13}C$ (‰, mPDB*)	σ_{SE}
Aspartic acid	Winter 22	-	-		7	-14	5	-		-
	Spring 21	35	13	25	13	-9	10	-10	-32	17
β-alanine	Winter 22	-	-		60	33	24	-		-
α-alanine	Winter 22				128	109	20			

Table S4.2: Position-specific carbon isotope ratios.

Delta values are reported both relative to our house amino acid standards as well as converted to our mPDB* reference frame. Reported errors are propagated σ_{SE}.

Specimen, Date of measurement	Compound	Fragment	Orbitrap params (AGC target, resolution, AQS window)	Valve turn timing	Peak integration timing	Background timing (if applicable)	Maximum NL score (NL score of background)
Spring 2021, 4/29/2021	Aspartic acid	156	$2*10^5$, 120,000, 153 to 160	18.05 to 19.6	18.7 to 24.5	n/a	$5*10^5$ ($5*10^3$)
Spring 2021, 4/30/2021	Aspartic acid	113	$5*10^4$, 120,000, 153 to 160	18.05 to 19.6	18.7 to 26.7	n/a	$2*10^5$ ($8*10^3$)
Spring 2021, 5/3/2021	Aspartic acid	198	$2*10^5$, 120,000, 153 to 160	18.05 to 19.6	18.7 to 25	n/a	$5*10^5$ ($5*10^3$)
Winter 2022, 1/31/2022	Aspartic acid	156	$2*10^5$, 60,000, 153.5 to 159.5	36.4 to 36.75	36.7 to 41.3	n/a	$1.5*10^5$ ($3.93*10^3$)]
Winter 2022, 2/3/2022	β-Alanine	126	$2*10^5$, 60,000, 123.5 to 129.5	19.3 to 20.2	19.7 to 39	19.4 to 19.55	$3.12*10^5$ ($1.14*10^4$)
Winter 2022, 2/3/2022	α-Alanine	140	$2*10^5$, 60,000, 137.5 to 144.5	13.6 to 14.3	13.99 to 26.3	n/a	$5*10^5$ ($8*10^3$)

Table S4.3: Experimental conditions for Spring 2021 and Winter 2022 specimens.

params = parameters.

	Retention time	Murchison, NL score	GSFC Blank, NL score	Caltech Blank, NL score
α-Alanine	13.74	$2.21*10^8$	$3.45*10^5$	$<1*10^4$
β-Alanine	19.61	$2*10^7$	$<1*10^4$	$<1*10^4$
Aspartic acid	36.9	$1.1*10^6$	$3*10^4$	$<1*10^4$

Table S4.4: Blanks.

Amino acid	$\delta^{13}C$ (‰, VPBD)	Reference
α-Alanine	+51.7±1.9 (D), +38.5±(2.2)	Pizzarello, Huang, and Fuller, 2004
	+38±10 (D), +40±19 (L)	Elsila, Charnley, et al., 2012
	+49±5 (D), +38±5 (L)	Glavin, Elsila, et al., 2020
	+25.5±3	Chimiak, Elsila, et al., 2021
	+30 (D), +27 (L)	Engel, Macko, and Silfer, 1990
	+52 (D), +38 (L)	Engel, Macko, and Silfer, 1990
Aspartic acid	+25.2 (D), -6.2 (L)	Pizzarello, Huang, and Fuller, 2004
	+4	Pizzarello, Krishnamurthy, et al., 1991
β-Alanine	+4.9±0.5	Pizzarello, Huang, and Fuller, 2004
	+10±6	Elsila, Charnley, et al., 2012
	+10±1	Glavin, Elsila, et al., 2020
	+5	Pizzarello, Cooper, and Flynn, 2006

Table S4.5: Prior compound specific carbon isotope values of Murchison amino acids.

Note that the reported range of values for each compound may reflect heterogeneities or some terrestrial contamination.

	Compound	$\delta^{13}C$ (‰, VPBD)	Reference
Murchison and Murchison parent body	Butyric acid	+11	Pizzarello, Cooper, and Flynn, 2006
	Carbon monoxide	-32	Pizzarello, Cooper, and Flynn, 2006
	Carbon dioxide	+29.1	Pizzarello, Cooper, and Flynn, 2006
	KCN	+5±3	Pizzarello, 2014
	Dicarboxylic and hydroxy di-carboxylic acid, Murchison	-6	Cronin et al., 1993
Cometary	HCN	+16 (+262, -172‰)	Cordiner et al., 2019
Interstellar	CO	+110 to +160‰	Chimiak, Elsila, et al., 2021; Lyons, Gharib-Nezhad, and Ayres, 2018

Table S4.6: Prior compound specific carbon isotope values of putative organic precursor and intermediate compounds.

Chapter 5

THE CARBON ISOTOPIC COMPOSITION OF ARCHEAN KEROGEN AND ITS RESILIENCE THROUGH THE ROCK CYCLE

Sarah S. Zeichner[1], Woodward W. Fischer[1], Noam Lotem[1], Kelsey R. Moore[1,2], Joshua E. Goldford[1], John M. Eiler[1]

Affiliations: [1]Division of Geological Planetary Sciences, California Institute of Technology, Pasadena, CA 91125, USA, [2]Department of Earth and Planetary Sciences, Johns Hopkins University, Baltimore, MD 21218 USA.

Introduction

The carbon cycle is a complex series of processes that governs the interactions between the biosphere and the geosphere, and has changed significantly since the formation of the Earth 4.56 billion years ago (Gya; Fig 5.1). Many of these differences reflect how distinct the Archean world was from the modern: The mantle was hotter, the partial pressure of atmospheric CO_2 was higher (perhaps as high as 1 bar), and there was little to no O_2 in the atmosphere and oceans (Schopf and Klein, 1992). The biosphere was nascent and early in its evolutionary development (Schopf, 1983). Prior to the evolution of oxygenic photosynthesis and the Great Oxygenation Event (GOE), it is thought that the productivity of the biosphere was limited (Ward, Rasmussen, and Fischer, 2019) and thus the flux of carbon between atmospheric CO_2 and organic carbon was lower than the modern (Des Marais, 2019). The GOE introduced a major shift in the global carbon cycle and a tremendous increase in global productivity (Fischer, Hemp, and Johnson, 2016), which has been observed in the rock record as an increase in the burial of organic carbon over time (Des Marais, 2019; Krissansen-Totton, Buick, and Catling, 2015).

The interpretation of buried, lithified, and metamorphosed refractory carbon has been a long-standing question for scientists interested in understanding putative biosignatures found in the oldest rocks on Earth (Hayes, Kaplan, and Wedeking, 1983) and interpreting them within the context of organic molecules from extraterrestrial samples, which present one form of an abiotic 'baseline' of organic chemical reactions from which life emerged. For instance, carbonaceous chondrites of similar estimated ages to Earth's have long been studied for their macromolecular carbonaceous material and high concentration of organic molecules (for example, carboxylic acids, ketones, amino acids; Elsila et al., 2012; Pizzarello et al., 1991; Sephton, 2002), whose near-racemic proportions and isotopic compositions provide support for an abiotic extraterrestrial synthesis mechanism (Chimiak et al., 2021; Elsila et al., 2012; Zeichner, Chimiak, et al., 2023). In contrast, the carbon cycle and C-rich organic matter within modern and Phanerozoic-age sedimentary rocks of all provenances is dominated by biologically generated organic matter (OM). Samples of organic matter from the Archean offer an opportunity to interrogate the transition from an abiotic Earth to a biotic one. However, all recognized Archean rocks have complex histories, and thus a detailed understanding of how organic matter evolves through the rock cycle is required to confidently determine whether or not Archean organic matter is biogenic.

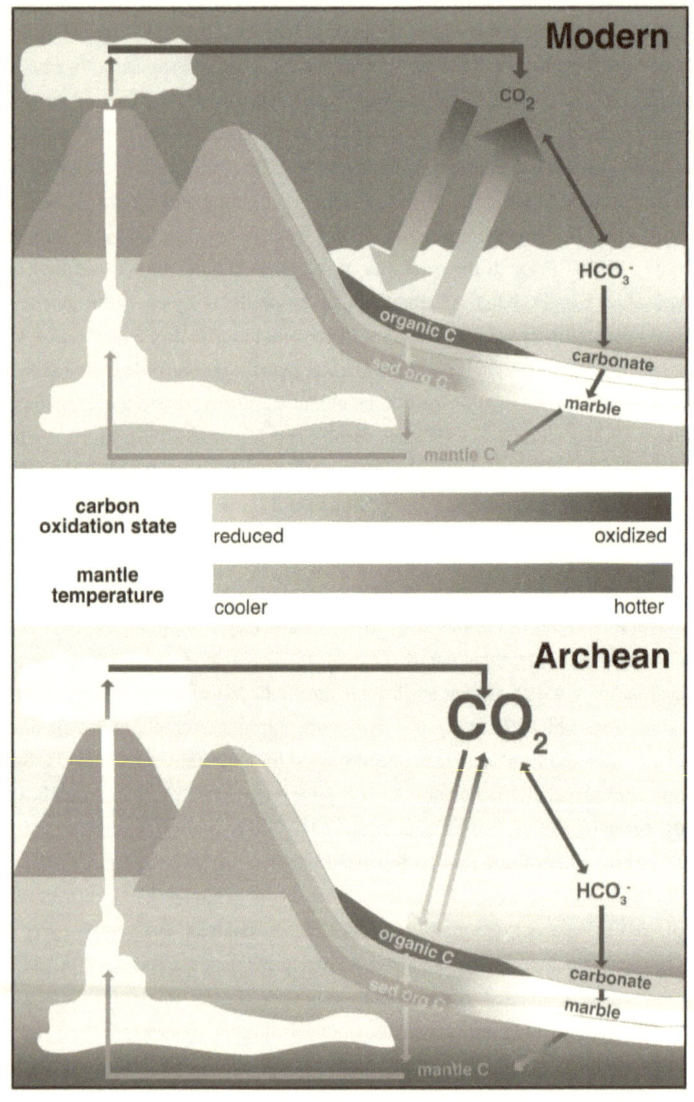

Figure 5.1: Idealized schematics of the carbon cycle, comparing modern and Archean.

Caption continued on next page.

Figure 5.1: Idealized schematics of the carbon cycle, comparing modern and Archean.

The carbon cycle regulates reservoirs and oxidation states of carbon on the Earth. The size of the arrows in this figure represent the fluxes for each step of the cycle, and are scaled to represent differences between the modern (top panel) and Archean (bottom panel) cycles, based on values reported in Des Marais, 2019. The arrow gradient colors represent changes in carbon oxidation states throughout the cycle, which is described briefly as follows: CO_2 is released into the atmosphere via volcanic activity. Atmospheric CO_2 equilibrates with bicarbonate (HCO_3^-) in the oceans, which precipitates as carbonate. Biology fixes CO_2 into organic carbon, which can then be further cycled by heterotrophs that derive energy from reduced carbon sources. In the modern carbon cycle, carbon fixation is driven by phototrophs and photosynthetic organisms, whose global proliferation drives the large modern fluxes in and out of the organic carbon pool. The Archean carbon cycle was distinct from that of the modern, due to a hotter mantle (redder), a more reduced atmosphere (more purple) with a higher partial pressure of CO_2, and lower rates of productivity. Distinct Archean alkalinity may have driven different rates of carbonate precipitation from the modern. In both modern and Archean carbon cycles, organic carbon gets incorporated into sedimentary rocks through burial and diagenesis. Org; organic, sed; sedimentary.

Ideally, studies of the Archean carbon cycle would target organic molecules that are chemically well-defined and diagnostic of a particular source, but the record of Archean organic matter is limited to kerogen. Kerogen is macromolecular carbon-rich material (Vandenbroucke and Largeau, 2007), insoluble in acid, base, or organic solvent, and thus typically operationally defined because it represents what is left when the solvent extractable organic compounds (bitumen) are removed. It is thought to be predominantly made up of degradation products of cell membranes, carotenoids, chlorophylls, wax esters, triglycerides, cross-linked proteins, and resins (Schopf and Klein, 1992; Vandenbroucke and Largeau, 2007), and classified into sub-types based on elemental ratios (that is, H/C and O/C), mineral associations, and whether or not the original biomolecular components are thought to have a marine, lacustrine or terrestrial origin (Vandenbroucke and Largeau, 2007). The origin of kerogen within Archean rocks has been a topic of longstanding debate, and because minimal molecular information is retained after biomass degrades to become kerogen, much of this debate focuses on interpretations of its carbon isotopic composition (Craig, 1954; Horita, 2005; Mojzsis et al., 1996; Rankama, 1954). Most of these studies operate under the premise that carbon isotope values are relatively unsusceptible to major shifts through the post-depositional processes that affect organic carbon in the rock cycle, and thus could provide clues into the metabolisms or abiogenic reactions that were dominant on early Earth. Indeed, in 1954, Kalervo Rankama argued that carbon isotope values substantially lower than the isotopic composition of the bulk-Earth—similar to the carbon isotopic composition of modern biomass—in Archean units must be evidence of life and thus can be used to interpret the Archean carbon isotope record as definitive evidence of early life on Earth (Rankama, 1954). Soon after, Harmon Craig challenged this uniformitarian argument, suggesting that the low carbon isotopic composition of refractory carbon does not necessarily require biology: These arguments questioned the use of low carbon isotope values to invoke biogenicity and highlighted that such assumptions underlie "unsolved problems in the geochemistry of carbon which must be investigated" (Craig, 1954).

Since the early Precambrian (meta)sedimentary rock record was first discovered and described, much of the research on the isotopic composition of the refractory carbon it contains has aimed to characterize the early biosphere through focused studies on the nuances of specific localities, lithologies, and time frames (for example, Brocks, Summons, et al., 2003; Eigenbrode and Freeman, 2006; French et al., 2015; Slotznick and Fischer, 2016; Ueno et al., 2002). Other studies have taken more holistic approaches by compiling and reinterpreting the carbon isotope measurements presented by previous studies (Des Marais, 2019; Hayes, Kaplan, and Wedeking, 1983; Krissansen-Totton, Buick, and Catling, 2015; Schopf and Klein, 1992). Characterizing the oldest refractory carbon on Earth has been difficult because of its overall scarcity, and understanding its origins in the context of the complex combination of processes that older OM has experienced. For instance, light carbon isotope values measured in carbonaceous inclusions of apatites from Isua and Akilia (3.8 Gya) were originally thought to be of a biological origin (Mojzsis et al., 1996). This finding was particularly valuable in the context of origins of life research; if the source rock was of sedimentary origin, its depositional environment could plausibly be interpreted as a submarine/deep marine environment whose ultramafic lithologies hint at the remnants of an Archean hydrothermal vent system. However, this interpretation of the Isua and Akilia apatites was challenged, both because the protolith may not be of sedimentary origin (and therefore OM found within these apatites could potentially derive from either an igneous or abiotic source; Fedo, 2000; Nutman et al., 1984), and because the same isotopic composition could be explained by Rayleigh distillation of abiogenic carbon (Eiler, Mojzsis, and Arrhenius, 1997). For even older and more altered samples, measurements of carbon isotope values of indigenous organic carbon are more challenging—for instance, many studies have measured and reported $\delta^{13}C$ values of graphite inclusions found within Hadean zircons, but only one measurement remains unchallenged as primary carbon rather than contamination (Bell et al., 2015).

The carbon isotopic composition of refractory carbon still represents an "unsolved problem in the geochemistry of carbon" (Craig, 1954). Both long-standing historical as well as more modern scientific questions highlight the need to revisit and enhance understanding of past studies of the carbon isotopic composition of refractory carbon from Archean rocks and carbonaceous chondrites. Not only could Archean kerogen potentially preserve a record of ancient biosignatures or the development of the ancient biosphere, there are also recent, ongoing and planned extraterrestrial sample return missions that aim to characterize material for its abiotic organic chemistry, and/or for signs of past or present life. In the samples collected by these studies, there may not be compounds that contain diagnostic molecular information that can ally them with specific organisms or processes; based on the meteorite record, we should expect that most of the OM that will be found within extraterrestrial samples will be macromolecular and insoluble. It is important to understand how to treat carbon isotope values of this OM, and how they may be modified with burial. While it is just as tempting now as it was at the dawn of carbon isotope geochemistry to invoke lower carbon isotope values of refractory carbon as definitive evidence of life (Green et al., 2021; Neveu et al., 2018; Rankama, 1954), applying this simple rubric to extraterrestrial samples or Earth's oldest rocks is fraught with uncertainties. Interpreting the isotopic composition of refractory carbon requires not only a consideration of the carbon cycle, biosphere, and geosphere of the modern, but also of post depositional effects of temperature, pressure, and degradation over time (that is, diagenesis, catagenesis, and metamorphosis; Hayes, Kaplan, and Wedeking, 1983).

Here, we present a study of three parts. First, we assembled a comprehensive dataset of carbon isotope values from prior measurements of the total organic carbon (TOC) and kerogen found within Archean rocks, which we analyzed across time and based on lithotype and metamorphic grade. Along with this compilation, we assembled descriptions of the mineralogy and fossils found in different Archean units. Second, we synthesized existing constraints on the evolution of organic carbonaceous matter to construct a mechanistic model of the evolution of its carbon isotopic composition from deposition through high-grade metamorphism. Finally, we applied this model to re-interpret particularly anomalous Archean organic carbon isotope records and to propose next steps for studies in this field.

A review of published constraints on the Archean carbon isotope record

Archean cratons and geobiological context

The geological record prior to the GOE is limited due to Earth's tectonics and rock cycle. Indeed, only 35 Archean cratons have been identified (Bleeker, 2003; Schopf and Klein, 1992), and these are thought to be derived from an even smaller number of supercratons (i.e. 3-5; Bleeker, 2003). The most well-studied formations containing Archean organic carbon are located in Western Australia, Greenland and South Africa, though there are also units in Canada, India, and Zimbabwe that have received lesser study (Fig. 5.2).

Along with the study of their carbon isotope records, Archean rocks have been investigated for potential sedimentological evidence of life as well as organic biomarker molecules. Putative stromatolites, carbonaceous spheroids, and biofilms have been identified in rocks of ~3.35 Gya (Strelley Pool Formation (Fm) of the Warrawoona group; Allwood, Grotzinger, et al., 2009; Marshall et al., 2007; Schopf, 1993, 2006), but the biogenicity of these structures remain debated by sedimentologists and paleontologists (for example, Brasier et al., 2002). A study of younger Archean rocks proposed the presence of organic biomarkers compounds in 2.78-2.45 Gya rocks (that is, steranes and hopanes; Brocks, Buick, et al., 2003), which would provide key evidence for the rise of oxygenic photosynthesis and a strikingly early evolution of eukaryotes. However, a similar measurement was replicated on a clean extraction of a nearby core, and the same result was not replicated (French et al., 2015). The earliest definitive molecular biomarkers are late Paleoproterozoic in age from the Barney Creek Fm (Brocks, Love, et al., 2005; Lepot, 2020; Love and Zumberge, 2021).

Biosignatures from Archean and Proterozoic basins are typically preserved in three major classes of sedimentary rocks: carbonate, chert, and shale or mudstone. Each of these rock types represents different depositional environments and taphonomic windows, both of which control preservation potential through the balancing act between degradation and preservation of organic compounds. Most of the best-preserved Precambrian microbial body fossils are preserved in chert, a crypto- to microcrystalline quartz that is thought to have precipitated rapidly in shallow marine or hydrothermal systems to preserve the fine detail of trapped cells as well as the kerogenous material from their cell walls and extra-polymeric substances (EPS). As a result of this rapid precipitation, cells trapped in chert become encased in micro- to-nanoscopic minerals smaller than a cell before extensive degradation can occur, allowing for the retention of cell shapes and preventing extensive degradation of the organic compounds. Most examples of body fossils are found in formations that post-date the GOE (Schopf and Klein, 1992). However, pre-GOE chert deposits from either potential hydrothermal settings (Fig Tree Formation; Onverwacht Group) or marine settings (Hamersley Group, Pongola Supergroup; Strelley Pool Chert; Panorama Formation; Apex Chert; Dresser Formation; Towers Formation; Pongola Supergroup; Donimalai Formation; Tumbiana Formation; Hamersely Group; Turee Creek Formationo; Lime Acres Formation; Gamohaan Group; Table S5.1) preserved clots of organic material as well as organic-rich microbial lamination. Some of these (for example, Gamohaan Group, Lime Acres Formation; Transvaal Basin; Turee Creek Group; Donimalai Formation) contain structures interpreted as putative microfossils (Altermann and Schopf, 1995; Altermann and Wotherspoon, 1995; Barlow and Kranendonk, 2018; Czaja, Beukes, and Osterhout, 2016; Fadel et al., 2017; Fischer, Schroeder, et al., 2009; Gandin, Wright, and Melezhik, 2005; Johnson et al., 2003; Klein, Beukes, and Schopf, 1987; Rasmussen and Muhling, 2023; Venkatachala et al., 1990; Waldbauer et al., 2009).

Carbonate deposits are frequently both spatially and temporally linked to chert in the Precambrian rock record. Many of the biosignature-hosting chert deposits occur as chert layers, nodules, and lenses within carbonate strata. In some of these, microbial structures like stromatolites are comprised of alternating layers of chert and carbonate (Altermann and Schopf, 1995; Altermann and Wotherspoon, 1995; Czaja, Beukes, and Osterhout, 2016; Fischer, Schroeder, et al., 2009; Flannery and Walter, 2012; Gandin, Wright, and Melezhik, 2005; Johnson et al., 2003; Klein, Beukes, and Schopf, 1987; Rivera and Sumner, 2014; Wright and Altermann, 2000). Like chert, many of these carbonates are finely crystalline and are thought to have precipitated relatively rapidly. As a result of this rapid and fine crystalline precipitation, they also preserve OM and microbial structures exceptionally well (for example,, Dresser Formation, Pongola Supergroup, Tumbiana Formation, Turee Greek, Lime Acres Formation, Gamohaan Group; Table S5.1). Unlike chert, however, actual body fossils are rarely found in carbonates. These biosignature-preserving chert and carbonate strata most commonly formed in shallow marine environments.

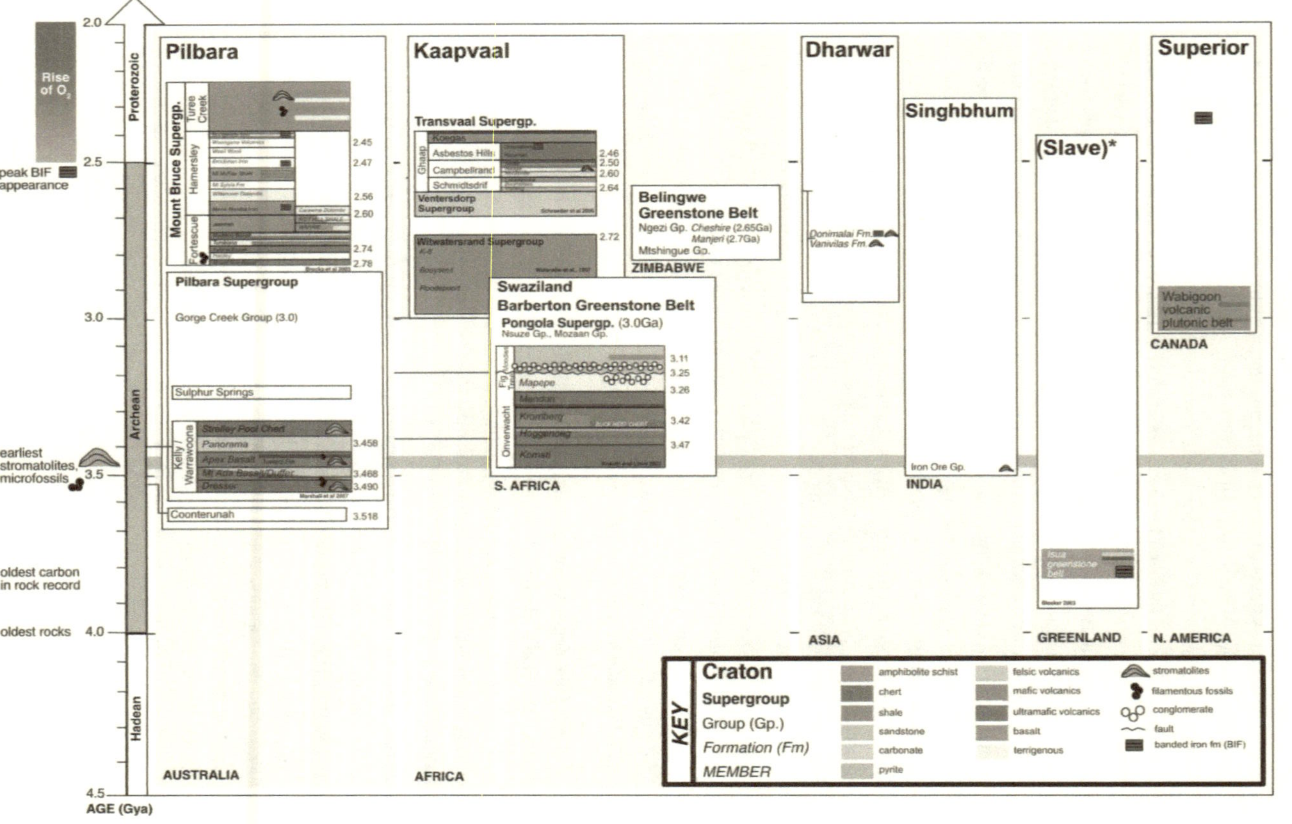

Figure 5.2: Archean units and predominant lithologies.

Caption continued on next page

Figure 5.2: Archean units and predominant lithologies.

Archean units and predominant lithologies. Due to the reworking of the rock cycle, there are a limited number of Archean cratons in the world today, with even a smaller subset of these including units with carbon preserved. Archean units relevant to studies of early biology and the evolution of life on Earth are presented, where width of unit represents length of time rather than width of actual stratigraphic units, and are placed on the geologic time scale to understand their relative time scales with respect to each other. Groups of rocks within each craton are subdivided into members, which are color coded by dominant lithology, with other lithological details represented as overlaid stripes or symbols. Approximate dates for each member are found to the right of lists of units. Sources for the stratigraphy and ages include Brocks, Buick, et al., 2003; Knauth and Lowe, 2003; Marshall et al., 2007; Schroder, Lacassie, and Beukes, 2006. See "A Note on Naming" [1] for information regarding the (Slave)* Craton. Fm.; formation, Gp.; group.

Shale and mudstone differ from chert and carbonate in several key ways. First, shale and mudstone from the Proterozoic and Archean may have formed in any number of environments from lagoons and lakes to deep marine environments. Additionally, the minerals that comprise these deposits are not predominantly comprised of authigenic precipitates. Instead, these deposits formed when mud, clay, and silt-sized sediments settled from suspension and were deposited and accumulated within low energy environments. Still, the small particle sizes of these sediments and the well-documented propensity for clay minerals to bind to functional groups of OM (Droppo et al., 1997; Lamb et al., 2020; Maggi, 2005; Zeichner, Ngheim, et al., 2021) can result in the trapping and binding of OM within shale and mudstone. In cases where the deposits formed in shallow lagoonal or marine settings, microbial communities that formed mats in these environments were preserved in the form of microbially laminated layers containing OM (for example, Moodies, Mozaan and Hamersley Groups; Table S5.1). For these cases, the OM would have been subject to cycles of production, alteration, and degradation similar to the OM preserved in chert and carbonate. In the case of deeper water deposits such as facies within the Moodies and Mozaan Groups (Table S5.1), it was more frequently the case that clay minerals bound to OM at one point in the transportation process and carried that OM to the sediment-water interface where it was deposited and preserved. The OM preserved in these deeper deposits, therefore, could represent any number of organic compounds produced by any number of organisms that lived either planktonically or in benthic mats and biofilms. Additionally, this OM likely experienced a range of degradation and alteration processes not only over the course of transportation but also after deposition where sediment-dwelling heterotrophs would have utilized the OM delivered to these environments as a source of reduced carbon. As is the case with chert- and carbonate-preserved OM, all of these degradation and preservation processes must be considered when interpreting the OM preserved in shale and mudstone. Several examples of organic-rich body fossils were preserved in shale deposits during the Proterozoic, but these all post-date the GOE and are thought to represent early eukaryotes including acritarchs and algae (Cohen and Macdonald, 2015; Javaux, Marshall, and Bekker, 2010; Knoll et al., 2006; Porter, 2004). For the purposes of this paper, we will focus on measurements of carbonaceous matter in pre-GOE deposits but not fossils, especially because the processes that drive the preservation of OM in sediments may be distinct from the processes that preserve fossils (Wiemann, Crawford, and Briggs, 2020).

Carbon isotope nomenclature, measurements and methods

We follow common nomenclature by quantifying carbon isotope ratios using "delta" notation, which reports the contrast between the $^{13}C/^{12}C$ ratio of sample and that same ratio in a reference standard, through the formula:

$$\delta^{13}C = (^{13/12}R_{sample}/^{13/12}R_{standard}) - 1 \qquad (5.1)$$

All data presented here are reported relative to the Vienna Pee Dee Belemnite carbon isotope standard (VPDB), and are reported in units of per-mille (‰; Brand et al., 2014). Any subscripts included with delta values serve as descriptors for the substrate (for example, $\delta^{13}C_{org}$ would represent the carbon isotopic composition of OM), rather than as an indicator of the standard reference frame. Most of the measurements of carbon isotope ratios compiled here were made by combusting samples in order to convert organic carbon to CO_2. These analyses have been mostly performed in recent years with an Elemental Analyzer, followed by isotopic analysis of the produced CO_2 using a gas source Isotope Ratio Mass Spectrometer.

Many papers report Precambrian carbon isotope values based on measurements of total organic carbon (TOC), which is the residual carbon within rocks that have been treated with strong acid to remove carbonate minerals. We distinguish TOC measurements from measurement of specifically targeted and extracted kerogen, which is prepared by removing all minerals and soluble organics from samples, and note in our analyses when we are discussing TOC versus kerogen. This is an important distinction because bulk rock TOC and kerogen may represent distinct organic phases because some components of TOC are likely to be intimately associated with the surfaces of minerals, and this may be excluded from a kerogen extract (for example, see Stueken et al., 2017, for an example of nitrogen isotope differences between bulk TOC and isolated kerogen). In our data compilation, we did not average any measurements for given localities and instead present the full range of carbon isotopic compositions measured. When referring to trends observed in the carbon isotopic composition of Archean organic carbon for both TOC or kerogen, we will refer to data as $\delta^{13}C_{org}$ values.

We compiled $\delta^{13}C_{org}$ values along with metadata for the geologic localities, dominant lithology of the units, % TOC (when available), and two different proxies for metamorphic grade. Refractory carbon is also sometimes characterized via RockEval, which measures, among other properties, the H/C ratio: The H/C ratio decreases as metamorphic grade increases. We use both H/C ratios and qualitative descriptions of metamorphic grade based on petrographic observations to interpret the effect of metamorphism on $\delta^{13}C_{org}$ values. The full data compilation, including data for each measurement, the original source of the data, locality and lithology information, available % TOC, H/C ratios and metamorphic grade are included in Zeichner, 2023. Data reduction and statistical analyses for this study and plots were performed and generated in RStudio, 2020.

Total organic carbon and kerogen carbon isotopic composition
$\delta^{13}C$ values of Archean TOC (n=2421) and kerogen (n=556) are plotted versus time in Fig 5.3A. To add context to our compilation and facilitate interpretation, we plotted our data alongside a previous compilation of $\delta^{13}C$ values of Archean carbonate-bearing rocks (n=731, Fig. 5.3A; where each data point may represent averages across several replicate measurements; Krissansen-Totton, Buick, and Catling, 2015). Archean TOC has a mean $\delta^{13}C$ of -30.5± 0.16‰ (1 standard error (SE); the standard deviation (SD) of individual measurements is ±8.0 ‰) and a median of -30.7 ‰; isolated kerogen from Archean rocks has a mean $\delta^{13}C$ of -33.7±0.48 (1 SE; the SD of individual measurements is ±11.3 ‰) and a median of -33.65 ‰. These populations are statistically significantly different from one another (based on two-sample Kolmogorov-Smirnov test (p-value < $2.2 * 10^{-16}$). For further comparison, analysis of a prior compilation of Phanerozoic TOC $\delta^{13}C$ values gave a mean $\delta^{13}C_{TOC}$ of -26.7±0.22 (1 SE for n = 449; the SD of individual measure-

ments is ±4.6‰) and a median of -27‰ (Krissansen-Totton, Buick, and Catling, 2015). Both the TOC and the kerogen carbon isotopic compositions of Archean samples compiled here were statistically significantly different from the carbon isotopic composition of the Phanerozoic TOC samples, based on a Kolmogorov-Smirnov test (p-values $< 2.2 \times 10^{-16}$). Together, these results demonstrated that the C isotopic composition of sedimentary rocks deposited in Archean basins is systematically and significantly lower than similar deposits generated over the past ~500 million years.

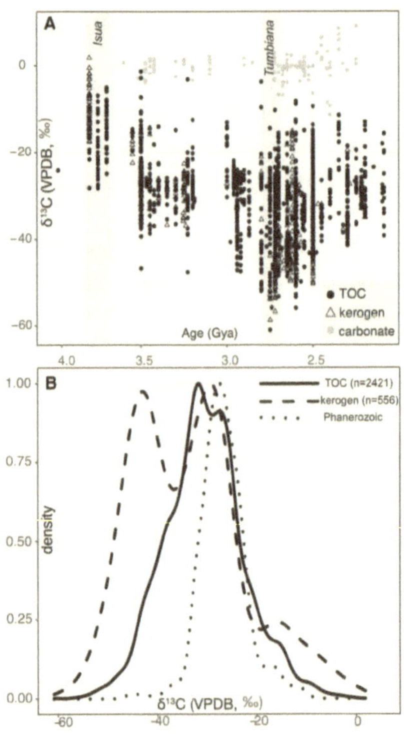

Figure 5.3: Organic carbon isotopes over time.

Caption continued on next page.

Figure 5.3: Organic carbon isotopes over time.

(A) Compiled $\delta^{13}C_{org}$ values for total organic carbon (TOC, filled circles) and and kerogen (open triangles) from rocks dating prior to the Great Oxygenation Event (GOE; black filled circles). References that originally presented the data within the compilation are listed in Zeichner, 2023. $\delta^{13}C_{carbonate}$ values from Krissansen-Totton, Buick, and Catling, 2015 are plotted for comparison (grey circles). ubsets of the data compilation that correspond to the Isua/Akilia and the Tumbiana formation outliers are highlighted with grey rectangles over the plot. (B) Kernel density estimates for $\delta^{13}C_{org}$ values for TOC (straight black line), kerogen (dashed line) and Phanerozoic organic carbon (dotted line; Krissansen-Totton, Buick, and Catling, 2015). TOC; total organic carbon.

To explore the differences between the $\delta^{13}C$ values of Archean TOC and Archean kerogen, as well as the difference between the isotopic compositions of both of these populations with that of Phanerozoic organic carbon, we generated kernel density estimates for the $\delta^{13}C$ values of Archean TOC, Archean kerogen and Phanerozoic TOC (Fig. 5.3B). $\delta^{13}C$ values of Archean OM are broadly much more variable than those of the Phanerozoic. This trend is reflected in both our calculated SDs (reported above) as well as in the kernel density estimates (Fig. 5.3B). While the average $\delta^{13}C$ value of Archean OM is lower than the average $\delta^{13}C$ value of Phanerozoic OM, that average value is representative of a spread, where some values are much lower (as low as $\sim$ -60‰) and some are much higher (up to -5‰). Outliers within the data set are concentrated within specific time frames, and correspond to specific localities/formations. For instance, high $\delta^{13}C$ values of -15 to 0‰ were found $\sim$3.8Gya within rocks from Isua and Akilia, while low $\delta^{13}C$ values down to -60‰ were found $\sim$2.7Gya within Tumbiana Fm rocks.

Further, the distribution of the Archean TOC $\delta^{13}C$ values was $\sim$unimodal (based on Hartigan's dip test; D = 0.007625, p-value = 0.4938), whereas the distribution of the Archean kerogen $\delta^{13}C$ values appeared to be bimodal (based on Hartigan's dip test; D = 0.040431, p-value = 1.946×10^{-6}). To our knowledge, this bimodality has not been identified by prior studies of the Archean carbon isotope record. These features may reflect differences in organic carbon genesis—for instance, dominant forms of either biotic or abiotic synthesis were different in the Archean than in the Phanerozoic—and/or variations in alternation/preservation. We subsampled the data in order to test whether multimodality within the kernel density estimate of distributions of $\delta^{13}C_{kerogen}$ was an artifact of preferential sampling (that is, specific localities known to have OM with anomalous isotopic compositions were sampled more; Fig. S5.2). Multimodality appears to be preserved even with subsampling which suggests that the spread in values is recording a real feature of the record (Fig. S5.2). This interpretation was supported by focused studies of specific formations from the Pilbara craton, where multimodal features with maxima at both higher (-10‰) and lower (-45‰) $\delta^{13}C$ values are preserved within samples with different dominant lithologies (Figs S5.2 and S5.3). We will re-address this multimodality in the section entitled "Reinterpreting Archean records."

Differences in carbon isotopic composition by lithology

Differential preservation of organic carbon within different lithologies could drive the variation in Archean $\delta^{13}C$ values. However, when averaged by eon, no clear patterns between lithology and carbon isotopic composition emerged (Fig. S5.1). Previous studies at the fm-scale highlighted small systematic differences in $\delta^{13}C_{org}$ values for distinct lithologies. For instance, studies have demonstrated that shale-hosted OM has lower $\delta^{13}C_{org}$ values than carbonate hosted OM for samples in both the Pilbara and Kaapvaal cratons (Eigenbrode and Freeman, 2006; Fischer, Schroeder, et al., 2009; Strauss and Beukes, 1996). This difference could be driven by actual differences in lithological preservation of OM, where the presence of dolomite and ankerite minerals may play some role in structural and isotopic evolution of OM, especially at high temperature where OM has the potential to undergo exchange with reactive inorganic C pools (detailed more in Sections "Model of carbon isotopic evolution through the rock cycle" and "Reinterpreting Archean records"; Chacko et al., 1991). However, higher $\delta^{13}C_{org}$ values associated with measurements of OM from carbonate-rich formations could also be explained by incomplete decarbonation prior to sample analysis (Fischer, Schroeder, et al., 2009), where $\delta^{13}C$ values of carbonate would drive the carbon isotope values of the measured sample to be higher.

We reanalyzed prior measurements of $\delta^{13}C_{org}$ from minimally metamorphosed (that is, greenschist facies and below) ~2.7 Ga units to probe the relationship between $\delta^{13}C_{org}$ values, lithology, and depositional environment (Fig. S5.3C). Based on our metanalyses of kernel density estimates of distributions of the $\delta^{13}C_{org}$ values from samples from the Hamersley basin, Pilbara Craton, carbonate and chert hosted OM had higher carbon isotope values than the values for organic phases hosted within other lithologies (including shale; Fig. S5.3C). For both the Hamerlsey and Fortescue Fms, chert-hosted OM had higher $\delta^{13}C_{org}$ values than OM hosted in other lithologies (Supplemental Fig. S5.3 B&C).

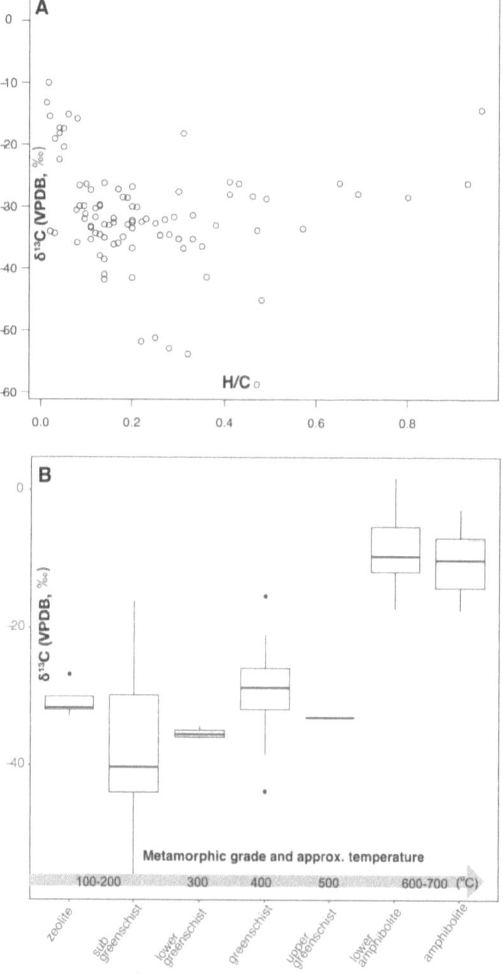

Figure 5.4: Metamorphic grade versus carbon isotopic composition of Archean carbon.

(A) Available H/C ratios versus $\delta^{13}C$ values for kerogen, which are often interpreted as a proxy for metamorphic grade. (B) Box-and-whisker plots for $\delta^{13}C$ values versus qualitative descriptors of metamorphic grade for each unit and approximate temperature for each metamorphic facies.

These higher values in chert-hosted OM may be driven by the fact that carbonate and chert are often co-associated, and that chert-associated samples often have low TOC contents; it more likely that incomplete decarbonation would affect the measured C isotope value of an organic-poor sample. We note that this trend in higher $\delta^{13}C_{org}$ values for chert and carbonate-hosted OM could not be extended to kernel density estimates for OM found within samples from the Tumbiana Fm (Fig. S5.3A). Additionally, the large range of $\delta^{13}C_{org}$ values within the Pilbara craton samples may make it challenging to interpret fm-scale trends based on lithology alone. In "Model of carbon isotopic evolution through the rock cycle" and "Reintepreting Archean records," we expound on the association between mineralogy, lithology, and post-depositional changes in structure and carbon isotopic composition, as well as the potential explanations for the range of $\delta^{13}C_{org}$ values of biogenic OM that could be reflected by the isotopic composition of the Pilbara samples.

Metamorphism and pre-GOE organic carbon isotopic composition

We also examined our data to understand how $\delta^{13}C_{org}$ values can change with metamorphism. We compared $\delta^{13}C$ values of Archean kerogen with two proxies for metamorphic grade: kerogen's H/C ratio and categorical description of metamorphic grade (Fig 5.4). As the H/C ratio decreased, we observed an enrichment in $\delta^{13}C$ values, particularly below H/C ratios of 0.2 (Fig 5.4A), as observed in previous studies (Hayes, Kaplan, and Wedeking, 1983). Likewise, we observed a shift in $\delta^{13}C$ values with metamorphic grade, although shifts of more than a few per-mille in the carbon isotopic composition of Archean kerogen were not observed until greenschist facies and above (Fig 5.4B). This observation is not new (Des Marais, 2019; Galvez et al., 2020; Schidlowski, 2001), but to our knowledge, no study has ever summarized this trend as we do here with a plot of $\delta^{13}C$ values versus qualitative description of metamorphic grade covering the full range of peak conditions. We focused our analysis on data from Archean kerogen rather than TOC because those studies included the most complete descriptions of metamorphic facies. Additionally, most studies measuring the $\delta^{13}C_{org}$ values of TOC lacked accompanying H/C ratios. To eliminate confusion, we grouped measurements from all sub-greenschist facies together.

H/C ratios decrease through catagenesis, metagenesis, and metamorphism due to the loss of functional groups and aromatization (see "Catagenesis/Metagenesis" and "Metamorphism" within "Model of carbon isotopic evolution through the rock cycle" below). A prior study presented an empirically derived, "corrective" polynomial to account for smaller shifts in H/C ratios at lower metamorphic grades, so that $\delta^{13}C$ values could be adjusted back to their "original values" based on measurements of this metamorphic proxy (Des Marais, 1997). However, corrections were only presented for $\delta^{13}C$ values measured from organic carbon deposited after the GOE, as there is limited kerogen data prior to the GOE — insufficient to generate a statistically meaningful polynomial fit. More importantly, interpreting this evolution as "correctable" by a polynomial fit conflates the effects of many reactions that could each impart a different and varying isotope effect. This isotopic fractionation would become more dramatic with increasing reaction progress, which might be expected to scale with temperature and pressure ("Model of carbon isotopic evolution through the rock cycle"). An improved understanding of such isotope effects is valuable for understanding the isotopic composition of all organic phases within rocks prior to the GOE, but becomes particularly important for understanding those of the oldest refractory carbon on Earth in rocks that have experienced high grades of metamorphism—particularly in rocks that predate robust sedimentological or paleontological evidence of life.

Model of carbon isotopic evolution through the rock cycle

Organic carbon within Archean rocks—even in the very best-preserved successions—has undergone a substantial amount of chemical change (Fig. 5.5), from low temperature degradation (Fig. 5.5A) to the production of oil and gas (Fig. 5.5B) to high temperature metamorphism (Fig. 5.5C). These changes span temperatures from 0 to >800°C (Fig 5.5C), and relate to the focus of several disparate disciplines—biology, low-temperature geochemistry, petroleum geoscience, and metamorphic petrology—each with their own vocabulary to describe processes involved in the maturation of organic matter. In some cases, there exists a gap where the focus of one field ends and the other begins. To address all these issues, we define nomenclature to describe the maturation of organic matter. We divide the evolution of organic matter into three 'alteration regimes': (i) Diagenesis is low-temperature alteration that occurs prior to and immediately following deposition (Fig. 5.5A); (ii) catagenesis and metagenesis refer to higher-temperature processes (50-200°C) that drive the production of oil and gas, respectively (Fig. 5.5B); and (iii) metamorphism refers to processes above 200°C that affect overmature organic matter (Fig. 5.5C).

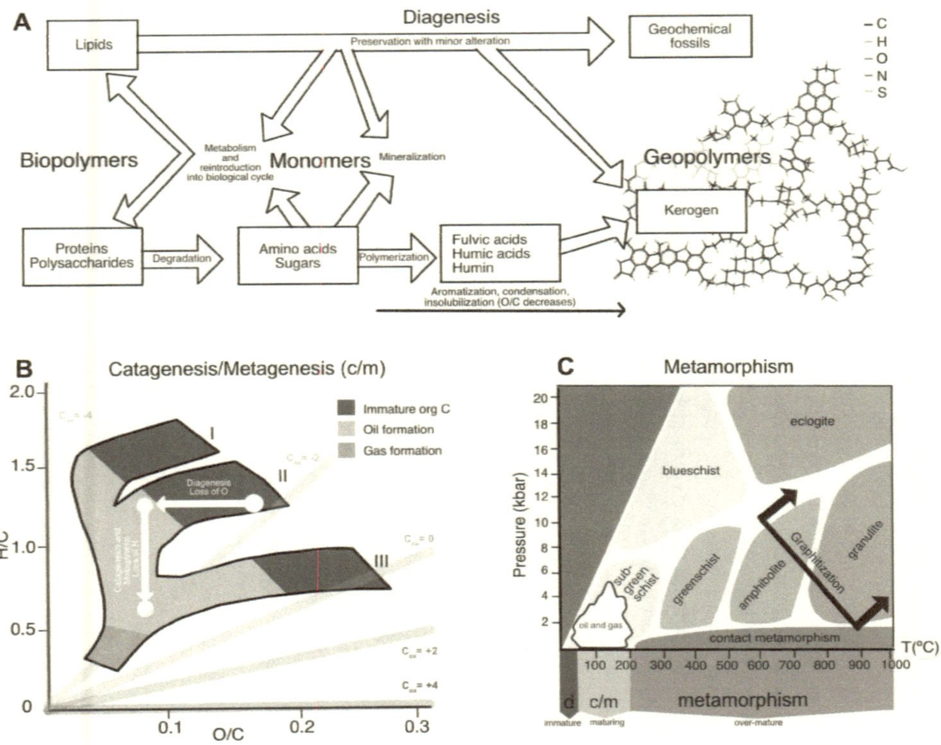

Figure 5.5: Alteration regimes.

Caption continued on next page

Isotope effects are caused by differences in thermodynamic stability and rates of reaction between isotopic forms of a compound. These effects are complex, and can impart a wide range of carbon isotopic fractionations at the scale of individual atomic positions (on the order of tens of in $\delta^{13}C$); depending on the strength of fractionation and proportions of strongly fractionated atomic sites, this can lead to observable shifts in the average isotopic composition of the compound or substrate. Kinetic isotope effects (KIEs) describe the isotopic fractionations associated with irreversible reactions, and represent the preference for one isotope over another during the rate limiting step of the reaction, whereas equilibrium isotope effects (EIEs) represent the partitioning of rare isotope between two or more materials when an exchangeable system has established equilibrium. Kinetic isotope effects are often reported as a ratio of reaction coefficients for the unsubstituted and substituted isotopologue (k_A/k_a, where A is the substituted mass of the isotope and a is the unsubstituted mass; for example, 13 and 12 for the substituted and unsubstituted isotopes of C). Both equilibrium and kinetic isotope effects can be quantified by "isotopic fractionation coefficients" (α values), which can be defined for both isotope ratios and delta values as follows:

$$\alpha_{i-j} = \frac{^{13/12}R_i}{^{13/12}R_j} = \frac{1000 + \delta^{13}C_i}{1000 + \delta^{13}C_j} \qquad (5.2)$$

where R's and δ's are the observed carbon isotope ratios and $\delta^{13}C$ values, respectively, for i and j, two distinct substrates of interest. Fractionation coefficients can be converted to epsilon values (ϵ), which are approximately equal to the difference (Δ) between the (δ) values of the two substrates:

$$\epsilon = 1000 \times (\alpha_{i-j} - 1) \approx \Delta = \delta^{13}C_i - \delta^{13}C_j \qquad (5.3)$$

Prior studies on the maturation of organic carbon have qualitatively described the effects of maturation reactions on the chemical structure of organic carbon. For instance, there have been studies quantifying the shifts in the G and D bands of Raman spectra as OM matures (Beyssac et al., 2002; Buseck and Beyssac, 2014). Changes in the chemical composition of kerogen through maturation have been characterized through pyrolysis experiments and through models (Clayton, 1991;

Figure 5.5: Alteration regimes.

(A) Deposition and diagenesis. Schematic of the transformation that biopolymers undergo following death and burial in sediments, as they are converted to monomers, fulvic and humic acids, and eventually humin and kerogen through diagenesis. Schematic was adapted from Tissot and Welte, 1984). (B) Van Krevelin diagram. Evolution of kerogen into oil and gas is often represented via a van Krevelin diagram, where cracking reactions are quantified for Type I (lacustrine), II (marine) and III (terrestrial) kerogens as loss of hydrogen and oxygen relative to the amount of carbon remaining. The van Krevelin diagram was modified from one published in Vandenbroucke and Largeau, 2007. Evolution of the carbon isotopic composition of immature-stage type II kerogen can be modeled simply as two distinct changes: first, net loss of oxygen in diagenesis through decarboxylation and dehydration (overall reduction of C), and second, net loss of hydrogen in catagenesis via cracking reactions (overall oxidation of C; Galvez et al., 2020). Purple and blue contour lines are overlain on the van Krevelin diagram to demonstrate how organic matter changes in oxidation state as it moves through the van Krevelin diagram (purple is more reduced, blue is more oxidized). (C) Metamorphism. A generalized scheme of metamorphic facies, based on figures in Bucher, 2005. Diagenetic, catagenetic/metagenetic (c/m), and metamorphic alteration regimes are highlighted in pressure and temperature space, along with the pressures and temperatures where graphitization begins. Each of these phases is labeled as "immature," "maturing," and "overmature," respectively, which is based on nomenclature in that organic geochemists use to refer to OM that has experienced varying degrees of thermal maturity.

Tocque et al., 2005). Likewise, prior studies on the $\delta^{13}C_{org}$ values of organic carbon have emphasized the complex range of processes that could be contributing to the isotopic composition of mature OM. However, we lack a mechanistic model for the evolution of the carbon isotopic composition of organic carbon through maturation and graphitization.

To understand the evolution of Archean kerogen, we focused on Type II OM, which is largely sapropelic, without a distinct shape or structure, and in the modern it is thought to be largely derived from phytoplankton (Tissot and Welte, 1984), with a starting composition of 50-60% amino acids, 40% carbohydrates, 5-30% lipids (Burdige, 2007). We chose to focus on Type II OM as we thought it would be more relevant to Archean organic carbon than are other types of phanerozoic kerogen.

We created a simplified model of organic matter and applied idealized reactions with defined isotope effects to track the change in the carbon isotopic composition of OM with elemental evolution (that is, loss of O, H and C). We isolated the range of processes affecting the structure and preservation of Type II OM into specific "alteration regimes" (Fig 5.5), which we defined as: synthesis, deposition and diagenesis (Fig 5.5A), catagenesis and metagenesis (Fig 5.5B) and metamorphism (Fig 5.5C). The van Krevelin diagram offers quantitative descriptions for how the H/C ratio (>1.4 in modern, living biomass) and O/C ratio co-evolve (Fig 5.5B) as each kerogen type moves from low temperature diagenesis through higher temperature catagenesis. In the van Krevelin diagram, these two alteration regimes can be deconstructed into two vectors: the loss of O+C during diagenesis (demonstrated by an arrow moving left on the diagram), and then the loss of H+C during catagenesis (demonstrated by an arrow moving down on the diagram; Fig 5.5B). We paired a model illustrating the isotope effects of primary reactions driving diagenetic and catagenetic elemental change, acting in series on the isotopic composition of idealized OM, with the change of the O/C and H/C ratios. OM following the diagenesis and catagenesis models was then subjected to exchange with reactive inorganic carbon (presumably within the metamorphic fluid in source rocks), as a model for C exchange during metamorphism.

Biological and abiological organic synthesis

Biological and abiological organic synthesis processes are known to impart characteristic and complex effects on the isotopic composition of carbon (and nitrogen, hydrogen, sulfur, and oxygen), which produces non-stochastic distributions of isotopes at both the site- and compound-specific level. In the modern carbon cycle, carbon stable isotope values of sedimentary organic matter vary locally depending on ecologically predominant carbon fixation metabolisms, water depth, CO_2 concentration, and other depositional and preservation biases that affect the carbon ultimately ending up in kerogen. Each of these reactions affects the ultimate carbon isotopic composition of sedimentary organic carbon, with the most well-known fractionation being the $\sim$-27‰ fractionation driven by the rate-limiting step of carbon fixation by RuBisCo within the Calvin-Benson-Bassham (CBB) cycle, also referred to as the reductive pentose phosphate (rPP) cycle. On the bulk scale, C3 plants tend to have lower $\delta^{13}C$ values that express the strong kinetic isotope effect of RuBisCo modulated by diffusion in and out of leaf stomata (overall ϵ = 10-22‰; Hayes, 2001), while the $\delta^{13}C$ values of C4 plants tend to be higher and more strongly dominated by transport into and through leaves (ϵ = 2-15‰; Hayes, 2001). Some metabolic pathways are even more fractionating — biomass synthesized via the reductive acetyl-CoA pathway can have $\delta^{13}C$ values as low as $\sim$-50 (ϵ

= 15-36‰; Hayes, 2001) — while other pathways impart little to no fractionation on biomass (for example, 3-hydroxypropionate cycle; Hayes, 2001).

It is known that both the molecular average and the intramolecular carbon isotopic composition of organic matter can vary. OM begins as several key "primary compound" classes, including proteins, nucleic acids, carbohydrates and lipids. Each primary compound has a distinct isotopic composition; for instance, glucose within cultured E.coli cells (average $\delta^{13}C \sim$ -10‰) was found to be enriched in its ^{13}C composition compared to lipids (average $\delta^{13}C \sim$ -17‰), while lipid acidic sites (average $\delta^{13}C \sim$ -13‰) have $\delta^{13}C$ values 20‰ higher than the neutral sites (average $\delta^{13}C \sim$ -35‰; Monson and Hayes, 1982). Broadly, Monson and Hayes suggested that this was because oxidized carbons within organic compounds preferentially incorporate ^{13}C, while reduced carbons preferentially incorporate ^{12}C. This inter- and intra-molecular isotopic partitioning affects the isotopic composition of preserved OM, especially due to preservation biases at high levels of OM remineralization. Further, prior work has demonstrated that organic carbon deposited and buried at different relative water depths can vary in its carbon isotopic composition (Eigenbrode and Freeman, 2006), which could be driven either by the dominant microbial metabolisms at different depths or variable preservation conditions. Due to evolution, one cannot necessarily assume uniformitarianism with regards to carbon isotopic fractionation (Flamholz et al., 2022; Wang et al., 2023), especially because the dominant Archean carbon fixation pathways are debated (Fuchs, 2011), the Archean environment was so different from today's, and some occurrences of Archean organic carbon may predate the evolution and proliferation of life on Earth.

Abiotic organic synthesis processes can also impart a range of distinctive isotopic fractionations, some of which may overlap with common "isotopic fingerprints" imparted by biological metabolisms. For instance, experimental Fischer Tropsch-type (FTT) synthesis studies have produced alkanes with $\delta^{13}C$ values ~-25 ‰ (McCollom, 2013; McCollom and Seewald, 2006). In contrast, studies fracture fluids of deep hydrothermal mines have measured $\delta^{13}C$ values of putative abiotic organic acids of ~-7 to -14‰ (Sherwood Lollar et al., 2021). For these two abiotic organic synthesis pathways alone, the carbon isotope values of organic matter overlap with the $\delta^{13}C$ values that can be produced by biological metabolisms described above. However, within kerogen that has lost structural information that could provide any context for its original source, $\delta^{13}C$ values still offer one of the primary and most valuable ways we may learn about carbon fixation within the early biosphere

– but only if we can learn to interpret these values correctly.

Deposition and diagenesis

The carbon isotopic composition of immature sedimentary OM is driven by two factors: what are the compounds (and their associated isotopic compositions) preferentially being preserved, and how do those compounds structurally and isotopically evolve through diagenesis? Among primary compounds, lipids are the least altered from death to preservation. In many cases at low temperature, even original lipid biomolecules are preserved, un- or little changed in chemical structure (that is, biomarkers; Fig 5.5A; Killops and Killops, 1990). However, only ~30% of the structure of OM within marine sediments can be identified as primary compounds Burdige, 2007, which suggests that there is significant remineralization of OM within particulate and dissolved organic carbon (POC and DOC). Sedimentary OM also may incorporate reworked kerogen and black carbon, or soot from burned biomass (Burdige, 2007), whose high-molecular weight carbon structure is thought to be formed by high temperature addition of acetylene (Mebel, Landera, and Kaiser, 2017).

We employ a "model compound" in our model of diagenesis to approximate the stoichiometry of immature OM and mass balance of leaving vs. remaining carbons: Decanoic acid is an 10-carbon fatty acid that represents the appropriate functional groups and elemental abundances characteristic of immature OM (most relevantly, decanoic acid has an O/C ratio of 0.2). This choice also considers an understanding of the controls on O/C ratios of primary compounds contributing to marine OM: Glucose, a representative sugar, has an O/C ratio of ~1; long chain fatty acids, representative of lipids, typically have O/C ratios of ~0.1; and amino acids, representative of protein, typically have an O/C ratio of ~0.3. Taken together, we estimated relative contributions of sugars, lipids and proteins to our model of 0.2, 0.6, and 0.2, respectively, which together contribute to an average O/C ratio of 0.2 and are consistent with both observations of preferential preservation of lipids over protein and sugars within the rock record and initial N/C ratios of immature organic matter, assuming the contribution of N is derived from protein (average N/C = 1/3; Wu et al., 2004). In our model, we assumed that 9 of the 10 carbons within decanoic acid had a $\delta^{13}C$ value of -37‰ and that the $\delta^{13}C$ value of the carboxyl site was -13‰, similar to the values reported in Monson and Hayes, but so that the average isotopic composition across all sites generated a $\delta^{13}C$ value similar to the average composition of Archean organic matter reported above (-34.6‰).

The structural and isotopic evolution of sedimentary OM is driven by its oxidation and remineralization. Broadly, oxidative stress is well known as a dominant driver of degradation following organism death. Both glycoxidation and lipoxidation reactions (for example, the Maillard reaction) have been mechanistically characterized in the context of food science studies (Vistoli et al., 2013), and applied to recent research on fossilization to understand how these reactions affect the preservation of molecules at the tissue-scale (Wiemann, Fabbri, et al., 2018). These reactions have also been investigated to understand the role of metal-catalysis in the chemical conversion and preservation of organic molecules (Moore et al., 2023). Qualitatively, exposure to and/or protection from biological and abiological oxidation reactions can affect the respective degradation and preservation of immature OM. For instance, the longer organics stay in the water column the more they may be susceptible to biological and abiological degradation (for example, via microbial reworking or Fenton chemistry that could break down organics; Garrido-Ramírez, Theng, and Mora, 2010). Conversely, the interaction of OM with minerals in sinking particles may help to enhance preservation potential: For instance, OM has been demonstrated to drive enhanced flocculation of clay minerals and thus enhanced organic deposition within both marine and terrestrial systems (Eisma, 1986; Zeichner, Ngheim, et al., 2021). Differences in oxygen exposure time, marine versus lacustrine environments, water depth, sediment flux and mineral precipitation rates, as well as the differences between the Archean and modern oceans (for example, higher Mn and Fe concentrations in Archean oceans could facilitate more OM preservation; Moore et al., 2023), would likely have effects on the preservation of organic carbon in the rock record. In modern environments, the majority of OM is thought to be remineralized (resulting in <1% burial efficiency of OM into long-term sedimentary storage), but it is likely that organic burial efficiency was higher in the Archean than in the modern due to anoxia (Burdige, 2007; Katsev and Crowe, 2015; Killops and Killops, 1990; Kipp, Lepland, and Buick, 2020).

The reactions that occur during diagenesis can be separated into reactions that only cause structural rearrangement of OM and those that remove C from the system. Rearrangement reactions that occur throughout OM degradation drive polymerization to help the OM achieve a more thermodynamically stable form. These reactions include: saturation of unsaturated bonds, irreversible rearrangements of Amadori products, condensation between amino acid residues and dicarbonyl groups, oxidative cross linking, cyclization, aromatization and isomerization (Killops and Killops,

1990; Vandenbroucke and Largeau, 2007; Vistoli et al., 2013). Despite innumerable structural changes imparted by these reactions (and, likely, isotope effects imparted at the site level, which up until this point remain largely unconstrained[2]) in the context of this study, we are interested in reactions that remove C from the system, how these reactions affect the ratio of C to other elements, and their primary isotope effects (as compared to reactions that cause structural transformations in organic molecules but not remove any carbons from the system and therefore do not drive shifts in the $\delta^{13}C$ value of bulk OM).

Decarboxylation is the primary carbon-fractionating reaction that occurs during diagenetic maturation of buried organic matter. This reaction is also one of the two main reactions responsible for the loss of oxygen from immature OM (along with dehydration or the loss of H_2O, note that this reaction does not remove C from the system and therefore is not fractionating for the purposes of this model). The loss of oxygen is clearly depicted within the van Krevelin diagram as a trend during early OM maturation, and drives an overall reduction in oxidation state of C (Fig 5.5B). The KIE on uncatalyzed decarboxylation reactions has been measured by prior studies: $k^{13}/k^{12} \sim 0.97$ for the CO_2 leaving the system (Bigeleisen and Friedman, 1949; Lewis et al., 1993; Lindsay, Bourns, and Thode, 1950; Marlier and O'Leary, 1984; O'Leary and Yapp, 1978).

We modeled the effect that decarboxylation and its primary C isotope effect has along paths captured by the van Krevelin diagram. Prior to the oil window, the O/C composition of Type II OM decreases from ~0.2 to 0.1, which in our model can be simplified as the decarboxylation of ~50% of the decanoic acid. We assumed that the decarboxylation would impart a KIE $(k^{13}/k^{12}) = 1.03$ on the staying carbon, selectively removing carbons with lower $\delta^{13}C$ values (relative to the initial value of the pool being removed) and initially driving the isotopic composition of the residue to be more ^{13}C-enriched. Thus, there are two opposing drivers on the $\delta^{13}C$ of residual carbon: Decarboxylation reaction acts on carboxyl sites that begin ^{13}C-rich compared to the isotopic composition of the other carbons (that is, -13‰ versus -37‰ for the other sites) but does so with a KIE that selects against [13]C. This competing effect can be seen in Fig 5.6A where the isotopic composition of the residue becomes more ^{13}C-enriched, and then ^{13}C deplete (note: if the decarboxylation reaction went to completion, the C isotopic composition of the final pool of OM would have a $\delta^{13}C$ value equal to that of the non-carboxyl sites of the decanoic acid that is, -37‰). However, the net effect of these processes is minimal: Our model predicted that the isotopic composition of the residue would remain similar to its original composition within 1‰ (Figure 5.6A&D; Supplemental Materials).

Other diagenetic reactions promote the loss or addition of other functional groups, as demonstrated by studies that have modeled the evolution of kerogen (Ungerer, Collell, and Yiannourakou, 2015). However, these reactions do not affect the number of carbon atoms within the system itself and so the isotope effects on the isotopic composition of residual carbon are likely to be negligible. For instance, sulfurization seems to have little effect on the overall $\delta^{13}C$ within average OM composition (although it may introduce site-specific isotope effects for the specific atoms undergoing sulfurization reactions; Putschew et al., 1998; Rosenberg, Kutuzov, and Amrani, 2018; Schouten et al., 1995). Deamination reactions impart quantifiable isotope effects on the leaving nitrogen as well as the C and H involved in the cleaved bond (Macko and Estep, 1984; Snider et al., 2002; Yu et al., 1981); we expect the order of isotope effect to be comparable to the isotopic fractionation of decarboxylation ($\sim$ -30‰) due to the similarities between the bonds between C-C and C-N. However, like for sulfurization, we would expect the overall effects on total OM isotopic composition to be negligible because these reactions do not remove C from the system.

Catagenesis/Metagenesis

Following the loss of carbon dioxide, further maturation of OM is characterized in the van Krevelin diagram as a loss of carbon and hydrogen through catagenesis and metagenesis. A prior model of catagenetic processes simplified the reactions driving catagenesis into a few key ones: homolytic cleavage, β-scission, H- abstraction, radical recombination, and radical isomerization (Xie, Formolo, and Eiler, 2022). Xie et al. argued that the carbon isotopic fractionations associated with catagenesis could be simplified and that the main fractionating reactions driving C isotopic change were homolytic cleavage and β-scission, which both impart isotope effects of k^{13}/k^{12} = 0.975 at 180°C on the reacting carbon of a first order bond (the exact KIE value varies by bond order and secondary/adjacent carbons demonstrate an isotope effect $\sim$ an order of magnitude less; Xie, Formolo, and Eiler, 2022).

We applied a similar approach the one used in Xie's study, but tracked the evolving isotopic composition of the residue rather than the products of catagenesis. We model the effects of homolytic cleavage and β-scission (which we refer to together as just "homolytic cleavage" for simplicity below) on a model residue. Our residue begins with an isotopic composition equal to composition of OM at the final stage of the diagenesis model ($\delta^{13}C$ = ~34.7‰). We modeled the chemical composition of the residue to have an H/C ratio that starts at 1.2 and evolves to 0.1. We note that the models of kerogen structure published by Ungerer describe immature Type II kerogen evolving from H/C ratios of 1.16 (Type II-A) to 0.58 (Type II-D); likewise, most studies that have investigated the isotopic composition of OM do not investigate the full range of H/C ratios that encompass alteration regimes from biomass to metamorphism. Here, we extend our model to encapsulate H/C ratios down to 0.1 to connect with the starting point of our metamorphism model (see "Metamorphism").

We simplified the homolytic cleavage model to remove only methane at each step (1C and 4H (implying that hydrogen abstraction and 'capping' converts primary radical products to stable alkanes). In reality the number of carbons lost with cleavage varies with phase of catagenesis, temperature, and gas formation mechanism (Milkov and Etiope, 2018; Milkov, Faiz, and Etiope, 2020; Seewald, Benitez-Nelson, and Whelan, 1998), but higher order carbon (C_{2+}) gases will remove more C at lower H/C ratios and with less fractionated ^{13}C, which will achieve a similar result to the sole removal of methane applied here (Fig 5.6B). As the H/C ratio evolved from 1.2 to 0.1, ^{13}C was distilled into the residues, driving an ~8‰ increase in the $\delta^{13}C$ value of residual organic matter (Fig 5.6B&D; Supplemental Materials and Methods). This result provides a mechanistic understanding for why the average carbon isotopic composition of kerogen becomes more positive during catagenesis. It also supports previously observed empirical trends between H/C ratio and the $\delta^{13}C$ value (Fig 5.6B replotted from Des Marais, 1997); note that the linear trend fit to our model matches the slope of Des Marais' data well, but would require different starting compositions of -33.3‰ and -36.5‰, as indicated by dotted dark teal lines). Perhaps most importantly, our model extends potential changes that can happen during catagenesis to encompass OM phases that range from fresh biomass to near-crystalline graphite, and span a large range of H/C ratios (that is, double the range in H/C ratio that is included in data within Des Marais, 1997). Our study demonstrates that the full range of catagenetic processes can fractionate up to 8‰, which is much more fractionating than suggested by prior studies who have suggested shifts of <5‰, thus implying that the carbon isotopic composition measured in sub-greenschist facies rocks provides a close approximation for that of synsedimentary environmental organic carbon.

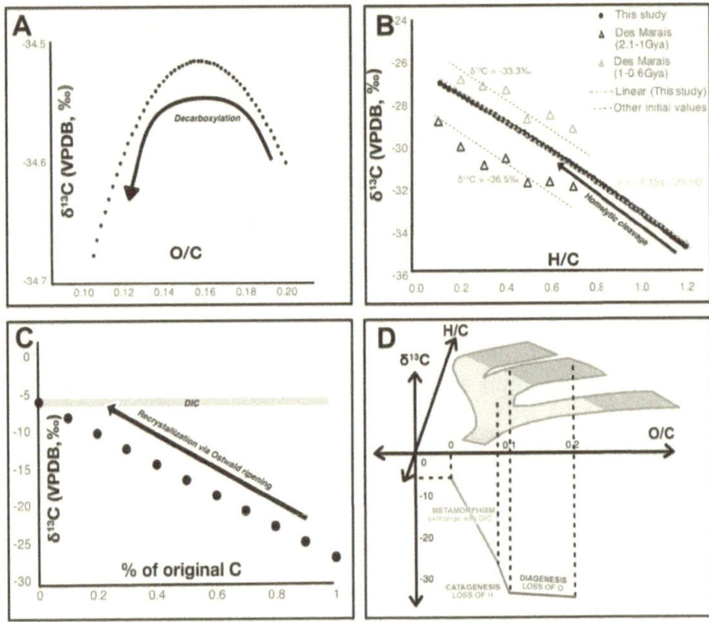

Figure 5.6: Model results showing changes in carbon isotopic composition as a function of maturation.

(A) Changes in $\delta^{13}C$ value of bulk OM during diagenesis are driven by decarboxylation, with an α on the OM residue of 0.97 (Marlier and O'Leary, 1984). (B) Changes in $\delta^{13}C$ value of bulk OM during catagenesis are driven by homolytic cleavage, with an α on the OM residue of 0.975 (Xie, Formolo, and Eiler, 2022). We fit a linear trend to our model, and equation is displayed on the figure. The same model, with different initial $\delta^{13}C$ values (dark teal dotted lines) fit prior compiled data (open grey and black triangles) presented in Des Marais, 1997. (C) Exchange with DIC during recrystallization of metamorphosed organic matter causes large positive shift of $\delta^{13}C$ value. (D) Evolutions in the $\delta^{13}C$ value versus evolutions along the van Krevelin diagram, viusalized as a 3-D plot.

Metamorphism

Finally, organic matter subject to high temperatures and pressures undergoes meteamorphism, which alters its structure and chemistry toward (though rarely reaching) pure crystalline graphite. At lower metamorphic grades (that is, greenschist facies and below), OM maturation towards graphitic structure is largely affected by condensation reactions such as aromatization (beginning at ~300 °C; Jing et al., 2007). However, as temperatures and pressures increase and rocks undergo amphibolite and granulite facies metamorphism, aromatic structures can be destroyed. C within OM at these temperatures and pressures undergoes exchange with the DIC in the surrounding fluid, before recrystallizing through Ostwald ripening as more highly-ordered graphite (Dunn and Valley, 1992). Previous studies of amphibolite and granulite facies marbles have shown that this process is accompanied by carbon isotope equilibration between organic matter and carbonate phases in the host rock ($\alpha_{carbonate-OM}$ = 1.006 at 600°C; presumably through dissolution of the smallest graphitic grains and growth of a smaller number of larger grains; Chacko et al., 1991; Dunn and Valley, 1992; Kitchen and Valley, 1995; Valley and O'Neil, 1981). However, the kinetics of carbon exchange with graphite are slow and equilibrium is unlikely to be obtained until peak metamorphic temperatures of 500-600°C (Valley and O'Neil, 1981). Thus, in the Archean samples included in this compilation, which generally reached peak metamorphic temperatures lower than this threshold, it is unlikely that any of the OM reached full equilibration, but rather, equilibrated partially with co-existing carbonate minerals and the DIC in the pore fluid. The expectation is that this will lead to substantially higher carbon isotope values with increasing metamorphic grade (Des Marais, 2019; Dunn and Valley, 1992; Galvez et al., 2020; Schidlowski, 2001).

To model this final stage of maturation, we modeled the exchange of a pool of OM with a starting $\delta^{13}C$ value equal to final $\delta^{13}C$ value for OM from the catagenesis model ($\delta^{13}C$ = -26.9‰), and a carbonate pool with an initial $\delta^{13}C$ value of 0‰ (a reasonable estimate for the $\delta^{13}C$ value of carbonate in Archean systems; Krissansen-Totton, Buick, and Catling, 2015). We assumed that the reactive inorganic C pool is infinitely large compared to the amount of OM in the system; with increased exchange and recrystallization of recalcitrant OM, the $\delta^{13}C$ value of equilibrated OM should approach the expected equilibrium isotopic composition between DIC and OM (estimated to have a $\delta^{13}C$ value = -6‰). This is obviously an oversimplification, but represents the average expected outcome, as DIC is abundant within seawater and pore fluids while sedimentary organic matter is more rare within the sedimentary and metamorphic rock record. At 20% exchange (80% of the original OM remains un-exchanged), the average $\delta^{13}C$ value of the OM was -22.7‰ (Fig. 5.6C&D; Supplemental Materials and Methods). Likewise, at 40, 60 and 80% exchange, the $\delta^{13}C$ values of the OM were -18.6‰, -14.4‰ and -10.2‰, respectively (Fig. 5.6C&D). This result is consistent with the $\delta^{13}C$ values of organic matter in high grade metamorphic samples from the Archean, and contextualizes these values within a mechanistic explanation of the shifting carbon isotopic composition associated with graphite grain growth during high grade metamorphism.

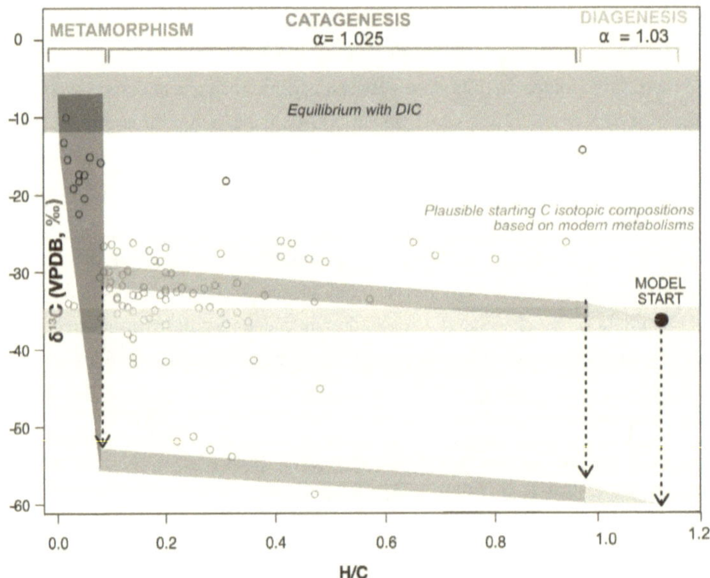

Figure 5.7: Model results applied to interpretation of carbon isotope record of kerogen versus H/C ratios.

Caption continued on next page.

Figure 5.7: Model results applied to interpretation of carbon isotope record of kerogen versus H/C ratios.

From an initial $\delta^{13}C$ value of -35‰ (black circle, darker purple bar) for decanoic acid neutral C, the carbon isotopic composition will evolve by ~1‰ through diagenesis and ~4‰ through catagenesis/metagenesis up until H/C ratios of 0.1. Metamorphism will drive overmature kerogen (H/C<0.1) towards isotopic equilibrium with DIC (~ -6‰), overlapping with many of the ^{13}C-enriched kerogen samples with values between -20‰ and -10‰ with H/C ratios <0.1. Values that are below that of isotopic equilibrium may reflect a partially equilibrated system, or a system where DIC and graphite underwent exchange at a lower temperature (Chacko et al., 1991). Modern biomass offers pathways for initial isotopic compositions to encompass a range of $\delta^{13}C$ values from ~-25‰ to -55‰ (light purple region, shifting the initial value down as indicated by dotted black arrows), and thus could explain a lot of kerogen $\delta^{13}C$ values found for moderately mature kerogen within that range. The two outlier values unconstrained by plausible ranges described above correspond to carbonate-rich environments with low TOC, and therefore may not necessarily derive from a biogenic source.

Reinterpreting Archean records

Our mechanistic model of the evolution of $\delta^{13}C_{org}$ values with changes in the O, H, and C ratios of OM through the rock cycle provides a framework to revisit the carbon isotopic composition of Archean organic matter (Fig 5.7). Major shifts in $\delta^{13}C$ values during end-stages of catagenesis and, especially, metagenesis through C exchange between OM and DIC can provide an explanation for ^{13}C-enriched values of amphibolite facies OM within Isua and Akilia samples. In contrast, our model gives no mechanism for substantial decreases of $\delta^{13}C$ values with maturation, meaning that exceptionally low $\delta^{13}C$ values observed in parts of the Archean record (particularly between 2.8 and 2.6 Gya) must derive from organic matter that was deposited with $\delta^{13}C$ values this low or lower. Given that this part of the geological record postdates generally accepted evidence for life, it is plausible that this interval marks a period dominated by contributions to sedimentary organic matter from metabolic pathways that generate exceptionally ^{13}C-depleted OM.

Isua, Akilia and Hadean zircons

The carbon isotopic composition of OM found within Isua and Akilia samples spans a range of values from a minimum of -27‰ to a maximum of -5‰ (Grassineau et al., 2006; Mojzsis et al., 1996; Rosing, 1999; Schopf and Klein, 1992; Shimoyama and Matsubaya, 1982; Tashiro et al., 2017; Ueno et al., 2002; Zuilen et al., 2003). Our model provides a mechanistic explanation for how biogenic organic matter could have reached the higher end of this distribution of carbon isotope values by exchanging with solid carbonates or fluid DIC. Indeed, to achieve the highest $\delta^{13}C$ values (~ -5‰) under amphibolite facies metamorphic conditions from a ^{13}C-deplete source, the OM would have to undergo full equilibration with reactive carbon in the surrounding fluid (Fig. 5.6&5.7). Likewise, moderately high values (~15‰) within these samples may be the result of biogenic OM that has undergone extreme catagenesis but not equilibration with reactive inorganic C. However, the idea that high $\delta^{13}C$ values may be measurements of completely abiotic carbon cannot be ruled out. Higher $\delta^{13}C$ values overlap with the carbon isotopic composition of some meteoritic organics (Glavin et al., 2018), or organics that could have been synthesized via abiotic organic synthesis processes (Sherwood Lollar et al., 2021).

Low $\delta^{13}C$ values within samples that have undergone amphibolite facies metamorphism could be the result of several potential scenarios. First, the OM characterized from these samples could be representative of original Archean biogenic OM that was prevented from equilibrating with the reactive carbon pool, either due to armoring by mantling minerals or fluid-absent metamorphism. Second, these values may represent rare grains that failed to dissolve, exchange, and re-precipitate as crystalline graphite. Finally, low $\delta^{13}C$ values may derive from contamination from more recent biogenic OM. Regardless of the explanation, we suggest that future studies focus on the textural and isotopic differences between re-crystallized graphite and more amorphous, overmature kerogen: Studies of kerogen that have not yet undergone crystallization and exchange, especially alongside measurements of the H/C ratios, can ensure that studies are targeting samples that have undergone the least amount of alteration and anchor that choice within the context of the van Krevelin diagram.

~ 2.7 Ga low $\delta^{13}C$ values

Based on our model, we argue that organic carbon that has not reached greenschist facies metamorphism will experience increases in its $\delta^{13}C$ value up to ~1‰ and ~8‰ due to diagenesis (O/C change from 0.2 to 0.1) and catagenesis (H/C change from 1.2 to 0.1), respectively. We note that the compounds themselves will undergo many structural changes (perhaps accompanied by much higher amplitude position-specific carbon isotope variations). However, there are no fractionations within our model that drive $\delta^{13}C_{org}$ values to be lower, which suggests that low values present within the record must have started out at or lower than their measured values.

The lowest carbon isotope values for Archean organic carbon occur within relatively un-metamorphosed units, such as the Tumbiana Formation (~2.7 Ga), which have measured $\delta^{13}C_{org}$ values as low as -60.9‰ (Slotznick and Fischer, 2016). These anomalously low values have made the Tumbiana Fm. a location of extensive past studies, yielding sufficient data for kernel density estimates of $\delta^{13}C_{kerogen}$ value distributions for specific formations in the Pilbara craton (Fig. S5.3). Kernel density estimates of $\delta^{13}C_{org}$ value distributions revealed higher values overall for data from the co-occurring Kaapvaal, Dhawar or Belingwe craton data, but with some multimodalities (Fig. S5.3). We note that within our data compilation, only three measurements of $\delta^{13}C_{kerogen}$ from the Tumbiana Fm are accompanied with H/C ratios, which range in their value from 0.28-0.47. Based on our catagenesis model, these H/C ratios could correspond to a maximum isotopic shift of 6‰ (assuming an initial H/C ratio of 1.2). Two samples from the Fortescue Fm (which has samples with similarly low values as those of the Tumbiana Fm with different multimodalities depending on lithology; Fig. S5.3B) have H/C ratios of 0.14 and 0.25, corresponding to $\delta^{13}C$ values of -41‰ and -51.2‰, respectively; these H/C ratios could at most correspond to a 7.6‰ shift. While additional measurements of H/C ratios are critical to better contextualizing these measurements within the context of potential isotopic changes during catagenesis, this potential shift up to ~8‰ in the C isotopic composition requires a discussion of the potential mechanisms that are capable of synthesizing biomass with low $\delta^{13}C$ values.

We argue that the distribution of $\delta^{13}C$ values in Archean kerogen is likely driven by variation in biological carbon fixation throughout Earth's history (that is, the multimodalities could correspond to distinct and competing dominant metabolisms that each impart their own diagnostic isotopic fingerprint on biomass). While there are currently seven known pathways capable of carbon fixation in the biosphere, there is no consensus for when these pathways emerged. Arguments based on a combination of comparative biology, phylogeny and physicochemical considerations have been used to propose an ordination of carbon fixation pathways throughout biological evolution. For example, it has been suggested that the reductive acetyl-CoA pathway was the first carbon fixation pathway, because it is present in both Archaea and Bacteria, is thermodynamically favorable, and uses potentially abundant geochemical reductants (for example, H_2; Weiss et al., 2016). More broadly, approaches purely based on comparative biology and phylogeny are unable to fully resolve and ordinate the emergence of different C fixation pathways due to processes like horizontal gene transfer, non-orthologous displacement, and organismal extinction.

A recent technique that explicitly models biochemical evolution through the use of network-based algorithms provides an alternative approach to exploring the history of metabolism (Goldford and Segré, 2018). For example, the network expansion algorithm models the emergence of metabolic networks through the recursive production of compounds from an initial starting set of compounds and reactions (Fig S5.4; Ebenhöh, Handorf, and Heinrich, 2004; Handorf, Ebenhöh, and Heinrich, 2005). The network expansion algorithm has been utilized in various studies to investigate significant questions about the origin and evolution of the biosphere. For instance, it was used to explore the impact of oxygen on the structure of metabolic networks (Raymond and Segrè, 2006) and to investigate geochemical constraints related to the origins of protometabolism (Goldford, Hartman, Marsland, et al., 2019; Goldford, Hartman, Smith, et al., 2017). A recent study proposed the potential of this algorithm to ordinate extant metabolic pathways throughout evolution, and demonstrated that—among extant biochemical metabolisms—the CBB cycle was the first of the carboxylation metabolisms to appear, followed by the reductive acetyl-CoA pathway (Fig. 5.8A; Goldford, Smith, et al., 2023) It is therefore notable that the trend in mean $\delta^{13}C$ of Archean organic carbon over time is consistent with this scenario: The $\delta^{13}C$ values of greenschist- and lower grade rocks start out with values similar to average carbon isotopic compositions produced by CBB cycle today (~-27‰) and then get lower, reaching a minimum of –60.9‰ near 2.7 Ga, before rising again toward values more typical of Proterozoic and Phanerozoic OM. Previous studies have suggested that the lower $\delta^{13}C$ values may either reflect anaerobic oxidation of methane (AOM) or the reductive acetyl Coenzyme A (CoA) pathway (Wood-Ljungdahl pathway; Slotznick and Fischer, 2016). Indeed, a prior study of AOM has measured $\delta^{13}C$ values of -133‰ (Machel, Krouse, and Sassen, 1995). Thus, this feature of the carbon isotope record in sub-greenschist facies Archean rocks may provide a benchmark for the evolutionary unfolding of carbon fixation in the biosphere.

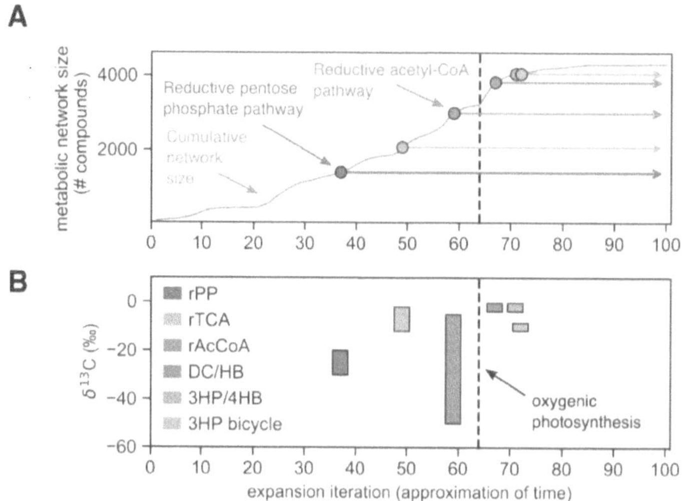

Figure 5.8: Timing the emergence of extant carboxylation pathways using network expansion.

(A) Cumulative network size (gray line, y-axis) after each iteration during network expansion (x-axis; Goldford, Smith, et al., 2023). Dots correspond to when all steps of the pathway become feasible, and remain feasible throughout the duration of the expansion process (arrows). (B) For 6/7 natural carbon fixation pathways, the ranges of $\delta^{13}C$ values from CO2 (Berg et al., 2010; Blaser, Dreisbach, and Conrad, 2013) are plotted on the y-axis. In both subplots A and B, the dotted black line corresponds to the expansion iteration when oxygenic photosynthesis emerges. rPP: Reductive pentose phosphate cycle, rTCA: Reductive citrate cycle, rAcCoA: Reductive acetyl-CoA pathway, DC/HB: Dicarboxylate-hydroxybutyrate cycle, 3HP/4HB: Hydroxypropionate-hydroxybutyrate cycle, 3HP bicycle: Hydroxypropionate bi-cycle.

Conclusions

The Archean carbon isotope record reflects a biosphere and a geosphere that were distinctly different from those of the modern. While some of the carbon isotope record may reflect abiotic and prebiotic organic syntheses (Estes et al., 2019; Fuchs, 2011; McCollom, 2013; Slotznick and Fischer, 2016; Sutherland, 2016), the majority of the record can be explained by the range of possible carbon isotopic compositions imparted by metabolisms, supported by recent network expansion algorithm models of metabolic processes, superimposed by structural changes and related isotopic fractionations through post-depositional degradation. Contextualizing $\delta^{13}C_{org}$ values with elemental (that is, H/C and O/C ratio measurements that place OM on the van Krevelin diagram) and structural studies (that is, the amorphicity of OM within samples that have experienced high metamorphic grades) can contribute to a stronger understanding of the OM's alteration history. These conclusions are aligned with conclusions drawn in past studies, but are supported by an understanding of elementary chemical steps to create an informed mechanistic isotopic model of OM maturation: Chemical isotope effects between oxidized and reduced, as well as condensed or vapor C species are large and potentially impactful for the isotopic composition of OM residues. Our model also implies that alteration processes impart strong position-specific isotopic fractionations within OM; advances in measurement capacities to this end may help to distinguish $\delta^{13}C_{org}$ values that could be explained by distinct formation/alteration processes. Improving our understanding in this area is particularly relevant now, as missions like Perseverance, OSIRIS-REx and Hayabusa2 store and return extraterrestrial material to Earth in order to study the organics, in the hopes of understanding potential extraterrestrial life, and improving our understanding of the potential formation of abiogenic organics on Earth, present and past.

[1]**A Note on Naming:** Names have power: It is valuable to consider how they acknowledge and perpetuate historical power dynamics. The craton that contains the Isua greenstone belt has been referred to by geologists as the "Slave Craton," which was named for the large lake in the region, "Great Slave Lake." "Great Slave" derives from the relationships between two First Peoples nations of the region: the Cree and the Dene. Historically, the Cree people referred to the Dene as "Slavey," and often raided and enslaved their neighbors to the north. There is an effort now to rename the lake for one of several of its traditional names, including: Tinde'e, Tucho or Big Lake. In the course of this work, we reached out to different representatives of the Dene Nation to discuss renaming efforts, but despite various conversations, were never able to find a conclusive answer regarding the current renaming efforts. To our knowledge, at the time this paper was submitted for publication, no traditional name has been chosen for the renaming of the lake: It is not our place to make this choice. Instead, we include the western name in parentheses to draw attention to complex history among peoples of the region and the importance of questioning and understanding naming conventions.

[2]Amination reactions involved in rearrangements of Amadori products are known to have an EIE of ~ 0 ‰ (Chimiak et al., 2021). Attempts to quantify the isotope effect of aromatization have been largely unsuccessful and produced ϵ values ranging from -1.2‰ to 2.7‰ (Bushnev, Burdel'naya, and Valyaeva, 2020; Freeman, Wakeham, and Hayes, 1994). Prior studies of isotope effects of isomerization of bicyclopentene to cyclopentadiene (Baldwin and Ghatlla, 1988) as well as the isomerization of carotenoids (Putschew et al., 1998) have not reported detectable isotope effects.

Funding sources: S.S.Z. was supported by the NSF Graduate Research Fellowship, the Simons Foundation, and the Resnick Sustainability Institute.

Author contributions: S.S.Z. conceived of the project, performed the literature review, compiled and analyzed the data, designed the mechanistic model, and wrote the paper. W.W.F. participated in investigation, data analysis, interpretation, and paper revisions. N.L. aided in designing the mechanistic model and collating the data compilation. K.R.M. compiled data related to taphonomy and sedimentology, and wrote the related sections. J.E.G. contributed interpretation with regards to biological implications of this study. J.M.E. participated in investigation, data analysis, interpretation, mechanistic model design, and paper revisions.

Data availability: The full data compilation for this study, as well as the spreadsheet associated with the maturation model can be found in an online repository. Additional materials and methods for the model, and supplemental table and figures are included in the Supplemental Materials and Methods at the end of the chapter.

References

Alleon, J. et al. (2018). "Chemical nature of the 3.4 Ga Strelley Pool microfossils". In: *Geochemical Perspectives Letters*, pp. 37–42. DOI: 10.7185/geochemlet.1817.

Allwood, A.C., J.P. Grotzinger, et al. (2009). "Controls on development and diversity of Early Archean stromatolites". In: *Proceedings of the National Academy of Sciences of the United States of America* 106.24, pp. 9548–9555. DOI: 10.1073/pnas.0903323106.

Allwood, A.C., B.S. Kamber, et al. (2010). "Trace elements record depositional history of an Early Archean stromatolitic carbonate platform". In: *Chemical Geology* 270.1–4, pp. 148–163. DOI: 10.1016/j.chemgeo.2009.11.013.

Allwood, A.C., M.R. Walter, B.S. Kamber, et al. (2006). "Stromatolite reef from the Early Archaean era of Australia". In: *Nature* 441.7094, pp. 714–718. DOI: 10.1038/nature04764.

Allwood, A.C., M.R. Walter, and C.P. Marshall (2006). "Raman spectroscopy reveals thermal palaeoenvironments of c.3.5 billion-year-old organic matter". In: *Vibrational Spectroscopy* 41.2, pp. 190–197. DOI: 10.1016/j.vibspec.2006.02.006.

Altermann, W. and J.W. Schopf (1995). "Microfossils from the Neoarchean Campbell Group, Griqualand West Sequence of the Transvaal Supergroup, and their paleoenvironmental and evolutionary implications". In: *Precambrian Research* 75.1–2, pp. 65–90. DOI: 10.1016/0301-9268(95)00018-Z.

Altermann, W. and J.McD Wotherspoon (1995). "The carbonates of the Transvaal and Griqualand West sequences of the Kaapvaal craton, with special reference to the Lime Acres limestone deposit". In: *Mineralium Deposita* 30.2. DOI: 10.1007/BF00189341.

Awramik, S.M. and H.P. Buchheim (2009). "A giant, Late Archean lake system: The Meentheena Member (Tumbiana Formation; Fortescue Group, Western Australia)". In: *Precambrian Research* 174.3–4, pp. 215–240. DOI: 10.1016/j.precamres.2009.07.005.

Baldwin, J. and N. Ghatlia (1988). "Deuterium and carbon-13 kinetic isotope effects for the reaction of OH with CH4". In: *The Journal of Chemical Physics* 99.5, pp. 3542–3552. DOI: 10.1063/1.466230.

Barlow, E.V. and M.J. Kranendonk (2018). "Snapshot of an early Paleoproterozoic ecosystem: Two diverse microfossil communities from the Turee Creek Group, Western Australia". In: *Geobiology* 16.5, pp. 449–475. DOI: 10.1111/gbi.12304.

Bell, E.A. et al. (2015). "Potentially biogenic carbon preserved in a 4.1 billion-year-old zircon". In: *Proceedings of the National Academy of Sciences of the United States of America* 112.47, pp. 14518–14521. DOI: 10.1073/pnas.1517557112.

Berg, I.A. et al. (2010). "Autotrophic carbon fixation in archaea". In: *Nature Reviews Microbiology* 8.6, pp. 447–460. DOI: `10.1038/nrmicro2365`.

Beukes, N.J. and D.R. Lowe (1989). "Environmental control on diverse stromatolite morphologies in the 3000 Myr Pongola Supergroup, South Africa". In: *Sedimentology* 36.3, pp. 383–397. DOI: `10.1111/j.1365-3091.1989.tb00615.x`.

Beyssac, O. et al. (2002). "Raman spectra of carbonaceous material in metasediments: A new geothermometer". In: *Journal of Metamorphic Geology* 20.9, pp. 859–871. DOI: `10.1046/j.1525-1314.2002.00408.x`.

Bigeleisen, J. and L. Friedman (1949). "C^{13} isotope effect in the decarboxylation of malonic acid". In: *The Journal of Chemical Physics* 17.ue 10, pp. 998–999. DOI: `10.1063/1.1747102`.

Blaser, M.B., L.K. Dreisbach, and R. Conrad (2013). "Carbon Isotope Fractionation of 11 Acetogenic Strains Grown on H_2 and CO_2". In: *Applied and Environmental Microbiology* 79.6, pp. 1787–1794. DOI: `10.1128/AEM.03203-12`.

Bleeker, W. (2003). "The late Archean record: A puzzle in ca. 35 pieces". In: *Lithos* 71.2–4, pp. 99–134. DOI: `10.1016/j.lithos.2003.07.003`.

Bolhar, R. and M. VanKranendonk (2007). "A non-marine depositional setting for the northern Fortescue Group, Pilbara Craton, inferred from trace element geochemistry of stromatolitic carbonates". In: *Precambrian Research* 155.3–4, pp. 229–250. DOI: `10.1016/j.precamres.2007.02.002`.

Bontognali, T.R.R., W.W. Fischer, and K.B. Föllmi (2013). "Siliciclastic associated banded iron formation from the 3.2Ga Moodies Group, Barberton Greenstone Belt, South Africa". In: *Precambrian Research* 226, pp. 116–124. DOI: `10.1016/j.precamres.2012.12.003`.

Brand, W.A. et al. (2014). "Assessment of international reference materials for isotope-ratio analysis (IUPAC technical report)". In: *Pure and Applied Chemistry* 86, pp. 425–467.

Brasier, M.D. et al. (2002). "Questioning the evidence for Earth's oldest fossils". In: *Nature* 416.6876, pp. 76–81. DOI: `10.1038/416076a`.

Brocks, J.J., R. Buick, et al. (2003). "Composition and syngeneity of molecular fossils from the 2.78 to 2.45 billion-year-old Mount Bruce Supergroup, Pilbara Craton, Western Australia". In: *Geochimica et Cosmochimica Acta* 67.22, pp. 4289–4319. DOI: `10.1016/S0016-7037(03)00208-4`.

Brocks, J.J., G.D. Love, et al. (2005). "Biomarker evidence for green and purple sulphur bacteria in a stratified Palaeoproterozoic sea". In: *Nature* 437.7060, pp. 866–870. DOI: `10.1038/nature04068`.

Brocks, J.J., R.E. Summons, et al. (2003). "Origin and significance of aromatic hydrocarbons in giant iron ore deposits of the late Archean Hamersley Basin, Western Australia". In: *Organic Geochemistry* 34.8, pp. 1161–1175. DOI: `10.1016/S0146-6380(03)00065-2`.

Bucher, K. (2005). "Metamorphic Rocks–Facies and Zones". In: *Encyclopedia of Geology*.

Burdige, D.J. (2007). "Preservation of organic matter in marine sediments: Controls, mechanisms, and an imbalance in sediment organic carbon budgets?" In: *Chemical Reviews* 107.2, pp. 467–485. DOI: 10.1021/cr050347q.

Buseck, P.R. and O. Beyssac (2014). "From organic matter to graphite: Graphitization". In: *Elements* 10.6, pp. 421–426. DOI: 10.2113/gselements.10.6.421.

Bushnev, D.A., N.S. Burdel'naya, and O.V. Valyaeva (2020). "$_{21}$ n-alkylbenzene and 1-n-alklylnaphthalene in Oils: Isotope Effect during Cyclization/Aromatization?" In: *Geochemistry International* 58.1, pp. 61–65. DOI: 10.1134/S0016702920010036.

Byerly, G.R. et al. (1996). "Prolonged magmatism and time constraints for sediment deposition in the early Archean Barberton greenstone belt: evidence from the Upper Onverwacht and Fig Tree groups". In: *Precambrian Research* 78.1–3, pp. 125–138. DOI: 10.1016/0301-9268(95)00073-9.

Chacko, T. et al. (1991). "Oxygen and carbon isotope fractionations between CO_2 and calcite". In: *Geochimica et Cosmochimica Acta* 55.10, pp. 2867–2882. DOI: 10.1016/0016-7037(91)90452-B.

Chimiak, L. et al. (2021). "Carbon isotope evidence for the substrates and mechanisms of prebiotic synthesis in the early solar system". In: *Geochimica et Cosmochimica Acta* 292, pp. 188–202. DOI: 10.1016/j.gca.2020.09.026.

Clayton, C. (1991). "Carbon isotope fractionation during natural gas generation from kerogen". In: *Marine and Petroleum Geology* 8, pp. 232–240.

Coffey, J.M. et al. (2013). "Sedimentology, stratigraphy and geochemistry of a stromatolite biofacies in the 2.72Ga Tumbiana Formation, Fortescue Group, Western Australia". In: *Precambrian Research* 236, pp. 282–296. DOI: 10.1016/j.precamres.2013.07.021.

Cohen, P.A. and F.A. Macdonald (2015). "The Proterozoic Record of Eukaryotes". In: *Paleobiology* 41.4, pp. 610–632. DOI: 10.1017/pab.2015.25.

Craig, H. (1954). "Geochemical implications of the isotopic composition of carbon in ancient rocks". In: *Geochimica et Cosmochimica Acta* 6, pp. 186–196. DOI: 10.1016/j.cossms.2010.07.001.

Czaja, A.D., N.J. Beukes, and J.T. Osterhout (2016). "Sulfur-oxidizing bacteria prior to the Great Oxidation Event from the 2.52 Ga Gamohaan Formation of South Africa". In: *Geology* 44.12, pp. 983–986. DOI: 10.1130/G38150.1.

Czaja, A.D., C.M. Johnson, et al. (2010). "Iron and carbon isotope evidence for ecosystem and environmental diversity in the ~2.7 to 2.5Ga Hamersley Province, Western Australia". In: *Earth and Planetary Science Letters* 292.1–2, pp. 170–180. DOI: 10.1016/j.epsl.2010.01.032.

Derenne, S. et al. (2008). "Molecular evidence for life in the 3.5 billion year old Warrawoona chert". In: *Earth and Planetary Science Letters* 272.1–2, pp. 476–480. DOI: 10.1016/j.epsl.2008.05.014.

Des Marais, D.J. (1997). "Isotopic evolution of the biogeochemical carbon cycle during the Proterozoic Eon". In: *Organic Geochemistry* 27.5–6, pp. 185–193.

– (2019). "Isotopic evolution of the biogeochemical carbon cycle during the precambrian". In: *Stable Isotope Geochemistry* 43, pp. 555–578. DOI: 10.1515/9781501508745-013.

Droppo, I.G. et al. (1997). "The Freshwater Floc: A Functional Relationship of Water and Organic and Inorganic Floc Constituents Affecting Suspended Sediment Properties". In: *The Interactions Between Sediments and Water: Proceedings of the 7th International Symposium.* Ed. by R.D. Evans, J. Wisniewski, and J.R. Wisniewski. Baveno, Italy: Springer Netherlands, pp. 43–53. DOI: 10.1007/978-94-011-5552-6_5.

Duda, J.-P. et al. (2016). "A Rare Glimpse of Paleoarchean Life: Geobiology of an Exceptionally Preserved Microbial Mat Facies from the 3.4 Ga Strelley Pool Formation, Western Australia". In: *PLOS ONE* 11.1, p. 0147629. DOI: 10.1371/journal.pone.0147629.

Dunn, S. and J.W. Valley (1992). "Calcite-graphite isotope thermometry: a test for polymetamorphism in marble, Tudor gabbro aureole, Ontario, Canada". In: *Journal of Metamorphic Geology* 10, pp. 487–501. DOI: 10.1016/j.cossms.2010.07.001.

Ebenhöh, O., T. Handorf, and R. Heinrich (2004). "Structural analysis of expanding metabolic networks". In: *Genome Informatics. International Conference on Genome Informatics* 15.1, pp. 35–45.

Eigenbrode, J.L. and K.H. Freeman (2006). "Late Archean rise of aerobic microbial ecosystems". In: *Proceedings of the National Academy of Sciences of the United States of America* 103.43, pp. 15759–15764. DOI: 10.1073/pnas.0607540103.

Eiler, J.M., S.J. Mojzsis, and G. Arrhenius (1997). "Carbon isotope evidence for early life". In: *Nature* 386.6626, p. 665.

Eisma, D. (1986). "Flocculation and de-flocculation of suspended matter in estuaries". In: *Netherlands Journal of Sea Research* 20.2–3, pp. 183–199. DOI: 10.1016/0077-7579(86)90041-4.

Elsila, J.E. et al. (2012). "Compound-specific carbon, nitrogen, and hydrogen isotopic ratios for amino acids in CM and CR chondrites and their use in evaluating potential formation pathways". In: *Meteoritics and Planetary Science* 47.9, pp. 1517–1536. DOI: 10.1111/j.1945-5100.2012.01415.x.

Estes, E.R. et al. (2019). "Abiotic synthesis of graphite in hydrothermal vents". In: *Nature Communications* 10.1. DOI: 10.1038/s41467-019-13216-z.

Fadel, A. et al. (2017). "Iron mineralization and taphonomy of microfossils of the 2.45–2.21 Ga Turee Creek Group, Western Australia". In: *Precambrian Research* 298, pp. 530–551. DOI: `10.1016/j.precamres.2017.07.003`.

Fedo, C.M. (2000). "Setting and origin for problematic rocks from the 3.7 Ga Isua Greenstone Belt, southern west Greenland: Earth's oldest coarse clastic sediments". In: *Precambrian Research*. Vol. 101.

Fischer, W.W., J. Hemp, and J.E. Johnson (2016). "Evolution of Oxygenic Photosynthesis". In: *Annual Review of Earth and Planetary Sciences* 44, pp. 647–683. DOI: `10.1146/annurev-earth-060313-054810`.

Fischer, W.W., S. Schroeder, et al. (2009). "Isotopic constraints on the Late Archean carbon cycle from the Transvaal Supergroup along the western margin of the Kaapvaal Craton, South Africa". In: *Precambrian Research* 169.1–4, pp. 15–27. DOI: `10.1016/j.precamres.2008.10.010`.

Flamholz, A.I. et al. (2022). "Trajectories for the evolution of bacterial CO2-concentrating mechanisms". In: *Proceedings of the National Academy of Sciences of the United States of America* 119.49. DOI: `10.1073/pnas.2210539119`.

Flannery, D.T., A.C. Allwood, et al. (2018). "Spatially-resolved isotopic study of carbon trapped in 3.43 Ga Strelley Pool Formation stromatolites". In: *Geochimica et Cosmochimica Acta* 223, pp. 21–35. DOI: `10.1016/j.gca.2017.11.028`.

Flannery, D.T. and M.R. Walter (2012). "Archean tufted microbial mats and the Great Oxidation Event: new insights into an ancient problem". In: *Australian Journal of Earth Sciences* 59.1, pp. 1–11. DOI: `10.1080/08120099.2011.607849`.

Freeman, K.H., S.G. Wakeham, and J.M. Hayes (1994). "Predictive isotopic biogeochemistry: Hydrocarbons from anoxic marine basins". In: *Organic Geochemistry* 21.6–7, pp. 629–644. DOI: `10.1016/0146-6380(94)90009-4`.

French, K.L. et al. (2015). "Reappraisal of hydrocarbon biomarkers in Archean rocks". In: *Proceedings of the National Academy of Sciences of the United States of America* 112.19, pp. 5915–5920. DOI: `10.1073/pnas.1419563112`.

Fuchs, G. (2011). "Alternative pathways of carbon dioxide fixation: Insights into the early evolution of life?" In: *Annual Review of Microbiology* 65. DOI: `10.1146/annurev-micro-090110-102801`.

Galvez, M.E. et al. (2020). "Materials and pathways of the organic carbon cycle through time". In: *Nature Geoscience* 13.August, pp. 535–546. DOI: `10.1038/s41561-020-0563-8`.

Gandin, A., D.T. Wright, and V. Melezhik (2005). "Vanished evaporites and carbonate formation in the Neoarchaean Kogelbeen and Gamohaan formations of the Campbellrand Subgroup, South Africa". In: *Journal of African Earth Sciences* 41.1–2, pp. 1–23. DOI: `10.1016/j.jafrearsci.2005.01.003`.

Garrido-Ramírez, E.G., B.K.G. Theng, and M.L. Mora (2010). "Clays and oxide minerals as catalysts and nanocatalysts in Fenton-like reactions - A review". In: *Applied Clay Science* 47.3–4, pp. 182–192. DOI: 10.1016/j.clay.2009.11.044.

Glavin, D.P. et al. (2018). "The origin and evolution of organic matter in carbonaceous chondrites and links to their parent bodies". In: *Primitive Meteorites and Asteroids. Elsevier*. Ed. by N. Abreu. DOI: 10.1016/B978-0-12-813325-5.00003-3.

Goldford, J.E., H. Hartman, R. Marsland, et al. (2019). "Environmental boundary conditions for the origin of life converge to an organo-sulfur metabolism". In: *Nature Ecology Evolution* 3.12, pp. 1715–1724. DOI: 10.1038/s41559-019-1018-8.

Goldford, J.E., H. Hartman, T.F. Smith, et al. (2017). "Remnants of an Ancient Metabolism without Phosphate". In: *Cell* 168.6, pp. 1126–1134 9. DOI: 10.1016/j.cell.2017.02.001.

Goldford, J.E. and D. Segré (2018). "Modern views of ancient metabolic networks". In: *Current Opinion in Systems Biology* 8, pp. 117–124. DOI: 10.1016/j.coisb.2018.01.004.

Goldford, J.E., H.B. Smith, et al. (2023). *Primitive purine biosynthesis connects ancient geochemistry 1 to modern metabolism 2 3*. BioRxiv. DOI: 10.1101/2022.10.07.511356.

Grassineau, N.V. et al. (2006). "Early life signatures in sulfur and carbon isotopes from Isua, Barberton, Wabigoon (Steep Rock), and Belingwe Greenstone Belts (3.8 to 2.7 Ga)". In: *Memoir of the Geological Society of America*, pp. 33–52. DOI: 10.1130/2006.1198(02).

Green, J. et al. (2021). "Call for a framework for reporting evidence for life beyond Earth". In: *Nature* 598.7882, pp. 575–579. DOI: 10.1038/s41586-021-03804-9.

Handorf, T., O. Ebenhöh, and R. Heinrich (2005). "Expanding Metabolic Networks: Scopes of Compounds, Robustness, and Evolution". In: *Journal of Molecular Evolution* 61.4, pp. 498–512. DOI: 10.1007/s00239-005-0027-1.

Hayes, J.M. (2001). "Fractionation of Carbon and Hydrogen Isotopes in Biosynthetic Processes". In: *Reviews in Mineralogy and Geochemistry* 43.1, pp. 225–277. DOI: 10.2138/gsrmg.43.1.225.

Hayes, J.M., I. Kaplan, and K. Wedeking (1983). "Precambrian Organic Geochemistry, Preservation of the Record". In: *The Earth's Earliest Biosphere: its origin and evolution*. Ed. by J.W. Schopf. Princeton University Press, pp. 93–134.

Heubeck, C. (2009). "An early ecosystem of Archean tidal microbial mats (Moodies Group, South Africa, ca. 3.2 Ga)". In: *Geology* 37.10, pp. 931–934. DOI: 10.1130/G30101A.1.

Heubeck, C. and D.R. Lowe (1994). "Depositional and tectonic setting of the Archean Moodies Group, Barberton Greenstone Belt, South Africa". In: *Precambrian Research* 68.3–4, pp. 257–290. DOI: 10.1016/0301-9268(94)90033-7.

Hofmann, A. and R. Bolhar (2007). "Carbonaceous Cherts in the Barberton Greenstone Belt and Their Significance for the Study of Early Life in the Archean Record". In: *Astrobiology* 7.2, pp. 355–388. DOI: 10.1089/ast.2005.0288.

Hofmann, H.J., K. Grey, et al. (1999). "Origin of 3.45 Ga coniform stromatolites in Warrawoona Group, Western Australia". In: *Geological Society of America Bulletin* 111.8, pp. 1256–1262. DOI: 10.1130/0016-7606(1999)111.

Homann, M., C. Heubeck, A. Airo, et al. (2015). "Morphological adaptations of 3.22 Ga-old tufted microbial mats to Archean coastal habitats (Moodies Group, Barberton Greenstone Belt, South Africa)." In: *Precambrian Research* 266, pp. 47–64. DOI: 10.1016/j.precamres.2015.04.018.

Homann, M., C. Heubeck, T.R.R. Bontognali, et al. (2016). "Evidence for cavity-dwelling microbial life in 3.22 Ga tidal deposits". In: *Geology* 44.1, pp. 51–54. DOI: 10.1130/G37272.1.

Horita, J. (2005). "Some perspectives on isotope biosignatures for early life". In: *Chemical Geology* 218.1-2 SPEC. ISS. Pp. 171–186. DOI: 10.1016/j.chemgeo.2005.01.017.

Javaux, E.J., A.H. Knoll, and M.R. Walter (2004). "TEM evidence for eukaryotic diversity in mid-Proterozoic oceans". In: *Geobiology* 2.3, pp. 121–132. DOI: 10.1111/j.1472-4677.2004.00027.x.

Javaux, E.J., C.P. Marshall, and A. Bekker (2010). "Organic-walled microfossils in 3.2-billion-year-old shallow-marine siliciclastic deposits". In: *Nature* 463.7283, pp. 934–938. DOI: 10.1038/nature08793.

Jing, M. et al. (2007). "Chemical structure evolution and mechanism during precarbonization of PAN-based stabilized fiber in the temperature range of 350-600°C". In: *Polymer Degradation and Stability* 92.9, pp. 1737–1742. DOI: 10.1016/j.polymdegradstab.2007.05.020.

Johnson, C.M. et al. (2003). "Ancient geochemical cycling in the earth as inferred from Fe isotope studies of banded iron formations from the Transvaal Craton". In: *Contributions to Mineralogy and Petrology* 144.5, pp. 523–547. DOI: 10.1007/s00410-002-0418-x.

Katsev, S. and S.A. Crowe (2015). "Organic carbon burial efficiencies in sediments: The power law of mineralization revisited". In: *Geology* 43.7, pp. 607–610. DOI: 10.1130/G36626.1.

Killops and Killops (1990). "Introduction to organic geochemistry". In: John Wiley Sons. Chap. Production, preservation and degradation of organic matter.

Kipp, M.A., A. Lepland, and R. Buick (2020). "Redox fluctuations, trace metal enrichment and phosphogenesis in the ~2.0 Ga Zaonega Formation". In: *Precambrian Research* 343, p. 105716. DOI: 10.1016/j.precamres.2020.105716.

Kitchen, N. and J.W. Valley (1995). "Carbon isotope thermometry in marbles of the Adirondack Mountains, New York". In: *Journal of Metamorphic Geology* 13, pp. 577–594.

Klein, C., N.J. Beukes, and J.W. Schopf (1987). "Filamentous microfossils in the early proterozoic transvaal supergroup: their morphology, significance, and paleoenvironmental setting". In: *Precambrian Research* 36.1, pp. 81–94. DOI: 10.1016/0301-9268(87)90018-0.

Knauth, L.P. and D.R. Lowe (2003). "High Archean climatic temperature inferred from oxygen isotope geochemistry of cherts in the 3.5 Ga Swaziland Supergroup, South Africa". In: *Bulletin of the Geological Society of America* 115.5, pp. 566–580. DOI: 10.1130/0016-7606(2003)115.

Knoll, A.H. et al. (2006). "Eukaryotic organisms in Proterozoic oceans". In: *Philosophical Transactions of the Royal Society B: Biological Sciences* 361.1470, pp. 1023–1038. DOI: 10.1098/rstb.2006.1843.

Kremer, B. and J. Kaźmierczak (2017). "Cellularly preserved microbial fossils from ~3.4 Ga deposits of South Africa: A testimony of early appearance of oxygenic life?" In: *Precambrian Research* 295, pp. 117–129. DOI: 10.1016/j.precamres.2017.04.023.

Krissansen-Totton, J., R. Buick, and D.C. Catling (2015). "A statistical analysis of the carbon isotope record from the Archean to phanerozoic and implications for the rise of oxygen". In: *American Journal of Science* 315.4, pp. 275–316. DOI: 10.2475/04.2015.01.

Lamb, M.P. et al. (2020). "Mud is transported as flocculated and suspended bed material". In: *Nature Geoscience* 13, pp. 566–570.

Lepot, K. (2020). "Signatures of early microbial life from the Archean (4 to 2.5 Ga) eon". In: *Earth-Science Reviews* 209.March, p. 103296. DOI: 10.1016/j.earscirev.2020.103296.

Lepot, K., K. Benzerara, G.E. Brown, et al. (2008). "Microbially influenced formation of 2.724-million-year-old stromatolites". In: *Nature Geoscience* 1.2, pp. 118–121. DOI: 10.1038/ngeo107.

Lepot, K., K. Benzerara, N. Rividi, et al. (2009). "Organic matter heterogeneities in 2.72Ga stromatolites: Alteration versus preservation by sulfur incorporation". In: *Geochimica et Cosmochimica Acta* 73.21, pp. 6579–6599. DOI: 10.1016/j.gca.2009.08.014.

Lepot, K., K.H. Williford, et al. (2013). "Texture-specific isotopic compositions in 3.4Gyr old organic matter support selective preservation in cell-like structures". In: *Geochimica et Cosmochimica Acta* 112, pp. 66–86. DOI: `10.1016/j.gca.2013.03.004`.

Lewis, C. et al. (1993). "Carbon Kinetic Isotope Effects on the Spontaneous and Antibody-Catalyzed Decarboxylation of 5-Nitro-3-carboxybenzisoxazole". In: *Journal of the American Chemical Society* 115.4, pp. 1410–1413. DOI: `10.1021/ja00057a025`.

Lindsay, J., A. Bourns, and H. Thode (1950). "C^{12} Isotope Effect in the Decarboyxlation of Normal Malonic Acid". In: *Canadia Journal of Chemistry* 29.2.

Love, G.D. and J.Alex Zumberge (2021). *Emerging patterns in proterozoic lipid biomarker records*. Cambridge University Press.

Lowe, D.R. (1983). "Restricted shallow-water sedimentation of Early Archean stromatolitic and evaporitic strata of the Strelley Pool Chert, Pilbara Block, Western Australia". In: *Precambrian Research* 19.3, pp. 239–283. DOI: `10.1016/0301-9268(83)90016-5`.

– (1994). "Abiological origin of described stromatolites older than 3.2 Ga". In: *Geology* 22.5, p. 387. DOI: `10.1130/0091-7613(1994)022`.

Machel, Hans G., H. Roy Krouse, and Roger Sassen (1995). "Products and distinguishing criteria of bacterial and thermochemical sulfate reduction". In: *Applied Geochemistry* 10 (4), pp. 373–389. DOI: `10.1016/0883-2927(95)00008-8`.

Macko, S.A. and M.L.F. Estep (1984). "Microbial alteration of stable nitrogen and carbon isotopic compositions of organic matter". In: *Organic Geochemistry* 6.C, pp. 787–790. DOI: `10.1016/0146-6380(84)90100-1`.

Maggi, F. (2005). "Flocculation dynamics of cohesive sediment". PhD book. Faculty of Civil Engineering and Geosciences, Delft University of Technology.

Marlier, J.F. and M.H. O'Leary (1984). "Carbon kinetic isotope effects o n the hydration of carbon dioxide and the dehydration of bicarbonate ion". In: *Journal of the American Chemical Society* 106.18, pp. 5054–5057. DOI: `10.1021/ja00330a003`.

Marshall, C.P. et al. (2007). "Structural characterization of kerogen in 3.4 Ga Archaean cherts from the Pilbara Craton, Western Australia". In: *Precambrian Research* 155.1–2, pp. 1–23. DOI: `10.1016/j.precamres.2006.12.014`.

Martindale, R.C. et al. (2015). "Sedimentology, chemostratigraphy, and stromatolites of lower Paleoproterozoic carbonates, Turee Creek Group, Western Australia". In: *Precambrian Research* 266, pp. 194–211. DOI: `10.1016/j.precamres.2015.05.021`.

McCollom, T.M. (2013). "Miller-Urey and Beyond: What Have We Learned About Prebiotic Organic Synthesis Reactions in the Past 60 Years?" In: *Annual Review of Earth and Planetary Sciences* 41.1, pp. 207–229. DOI: 10.1146/annurev-earth-040610-133457.

McCollom, T.M. and J.S. Seewald (2006). "Carbon isotope composition of organic compounds produced by abiotic synthesis under hydrothermal conditions". In: *Earth and Planetary Science Letters* 243.1–2, pp. 74–84. DOI: 10.1016/j.epsl.2006.01.027.

Mebel, A.M., A. Landera, and R.I. Kaiser (2017). "Formation Mechanisms of Naphthalene and Indene: From the Interstellar Medium to Combustion Flames". In: *Journal of Physical Chemistry A* 121.5, pp. 901–926. DOI: 10.1021/acs.jpca.6b09735.

Milkov, A.V. and G. Etiope (2018). "Revised genetic diagrams for natural gases based on a global dataset of >20,000 samples". In: *Organic Geochemistry* 125, pp. 109–120. DOI: 10.1016/j.orggeochem.2018.09.002.

Milkov, A.V., M. Faiz, and G. Etiope (2020). "Geochemistry of shale gases from around the world: Composition, origins, isotope reversals and rollovers, and implications for the exploration of shale plays". In: *Organic Geochemistry* 143. DOI: 10.1016/j.orggeochem.2020.103997.

Mojzsis, S.J. et al. (1996). "Evidence for life on Earth before 3,800 million years ago". In: *Nature* 384.6604, pp. 55–59. DOI: 10.1038/384055a0.

Monson, K.D. and J.M. Hayes (1982). "Carbon isotopic fractionation in the biosynthesis of bacterial fatty acids. Ozonolysis of unsaturated fatty acids as a means of determining the intramolecular distribution of carbon isotopes". In: *Geochimica et Cosmochimica Acta* 46.2, pp. 139–149. DOI: 10.1016/0016-7037(82)90241-1.

Moore, O.W. et al. (2023). "Long-term organic carbon preservation enhanced by iron and manganese". In: *Nature*. DOI: 10.1038/s41586-023-06325-9.

Morag, N. et al. (2016). "Microstructure-specific carbon isotopic signatures of organic matter from ~3.5 Ga cherts of the Pilbara Craton support a biologic origin". In: *Precambrian Research* 275, pp. 429–449. DOI: 10.1016/j.precamres.2016.01.014.

Muir, M.D. and D.O. Hall (1974). "Diverse microfossils in Precambrian Onverwacht group rocks of South Africa". In: *Nature* 252.5482, pp. 376–378. DOI: 10.1038/252376a0.

Murphy, R.J. et al. (2016). "Complex patterns in fossilized stromatolites revealed by hyperspectral imaging (400-2496 nm)". In: *Geobiology* 14.5, pp. 419–439. DOI: 10.1111/gbi.12184.

Neveu, M. et al. (2018). "The ladder of life detection". In: *Astrobiology* 18.11, pp. 1375–1402. DOI: 10.1089/ast.2017.1773.

Noffke, N., K.A. Eriksson, et al. (2006). "A new window into Early Archean life: Microbial mats in Earth's oldest siliciclastic tidal deposits (3.2 Ga Moodies Group, South Africa)". In: *Geology* 34.4, p. 253. DOI: 10.1130/G22246.1.

Noffke, N., R. Hazen, and N. Nhleko (2003). "Earth's earliest microbial mats in a siliciclastic marine environment (2.9 Ga Mozaan Group, Sout Africa)". In: *Geology* 31.8, p. 673. DOI: 10.1130/G19704.1.

Nutman, A.P. et al. (1984). "Stratigraphic and geochemical evidence for the depositional environment of the early Archaean Isua supracrustal belt, southern west Greenland". In: *Precambrian Research* 25.4, pp. 365–396. DOI: 10.1016/0301-9268(84)90010-X.

O'Leary, M.H. and C.J. Yapp (1978). "Equilibrium carbon isotope effect on a decarboxylation reaction". In: *Biochemical and Biophysical Research Communications* 80.1, pp. 155–160.

Ono, S. (2006). "Early evolution of atmospheric oxygen from multiple-sulfur and carbon isotope records of the 2.9 Ga Mozaan Group of the Pongola Supergroup, Southern Africa". In: *South African Journal of Geology* 109.1–2, pp. 97–108. DOI: 10.2113/gssajg.109.1-2.97.

Ono, S. et al. (2003). "New insights into Archean sulfur cycle from mass-independent sulfur isotope records from the Hamersley Basin, Australia". In: *Earth and Planetary Science Letters* 213.1–2, pp. 15–30. DOI: 10.1016/S0012-821X(03)00295-4.

Partridge, M.A. et al. (2008). "Pyrite paragenesis and multiple sulfur isotope distribution in late Archean and early Paleoproterozoic Hamersley Basin sediments". In: *Earth and Planetary Science Letters* 272.1–2, pp. 41–49. DOI: 10.1016/j.epsl.2008.03.051.

Pizzarello, S. et al. (1991). "Isotopic analyses of amino acids from the Murchison meteorite". In: *Geochimica et Cosmochimica Acta* 55.3, pp. 905–910. DOI: 10.1016/0016-7037(91)90350-E.

Porter, S.M. (2004). "The fossil record of early eukaryotic diversification". In: *The Paleontological Society Papers* 10, pp. 35–50. DOI: 10.1017/S1089332600002321.

Putschew, A. et al. (1998). "Carbon isotope characteristics of the diaromatic carotenoid, isorenieratene (intact and sulfide-bound) and a novel isomer in sediments". In: *Organic Geochemistry* 29.8, pp. 1849–1856. DOI: 10.1016/S0146-6380(98)00170-3.

Rankama, K. (1954). "The isotopic constitution of carbon in ancient rocks as an indicator of its biogenic or nonbiogenic origin". In: *Geochimica et Cosmochimica Acta* 5, pp. 142–152. DOI: 10.1016/j.cossms.2010.07.001.

Rasmussen, B. and J.R. Muhling (2023). "Organic carbon generation in 3.5-billion-year-old basalt-hosted seafloor hydrothermal vent systems". In: *Science Advances* 9.5. DOI: 10.1126/sciadv.add7925.

Raymond, J. and D. Segrè (2006). "The Effect of Oxygen on Biochemical Networks and the Evolution of Complex Life". In: *Science* 311.5768, pp. 1764–1767. DOI: 10.1126/science.1118439.

Rivera, M.J. and D.Y. Sumner (2014). "Unraveling the three-dimensional morphology of Archean microbialites". In: *Journal of Paleontology* 88.4, pp. 719–726. DOI: 10.1666/13-084.

Rosenberg, Y.O., I. Kutuzov, and A. Amrani (2018). "Sulfurization as a preservation mechanism for the $\delta^{13}C$ of biomarkers". In: *Organic Geochemistry* 125, pp. 66–69. DOI: 10.1016/j.orggeochem.2018.08.010.

Rosing, M.T. (1999). "^{13}C-depleted carbon microparticles in ≥3700-Ma sea-floor sedimentary rocks from west Greenland". In: *Science* 283.5402, pp. 674–676. DOI: 10.1126/science.283.5402.674.

RStudio (2020). "RStudio: Integrated Development for R". In: *RStudio, Inc.* p. Boston, MA.

Sakurai, R. et al. (2005). "Facies architecture and sequence-stratigraphic features of the Tumbiana Formation in the Pilbara Craton, northwestern Australia: Implications for depositional environments of oxygenic stromatolites during the Late Archean". In: *Precambrian Research* 138.3–4, pp. 255–273. DOI: 10.1016/j.precamres.2005.05.008.

Schidlowski, M. (2001). "Carbon isotopes as biogeochemical recorders of life over 3.8 Ga of earth history: Evolution of a concept". In: *Precambrian Research* 106.1–2, pp. 117–134. DOI: 10.1016/S0301-9268(00)00128-5.

Schopf, J.W. (1983). *Earth's earliest biosphere: Its origin and evolution.*

– (1993). "Microfossils of the early Archean apex chert: New evidence of the antiquity of life". In: *Science* 260.5108, pp. 640–646. DOI: 10.1126/science.260.5108.640.

– (2006). "Fossil evidence of Archaean life". In: *Philosophical Transactions of the Royal Society B: Biological Sciences* 361.1470, pp. 869–885. DOI: 10.1098/rstb.2006.1834.

Schopf, J.W. and C. Klein (1992). *The Proterozoic Biosphere.*

Schopf, J.W. and B.M. Packer (1987). "Early Archean (3.3-Billion to 3.5-Billion-Year-Old) Microfossils from Warrawoona Group, Australia". In: *Science* 237.4810, pp. 70–73. DOI: 10.1126/science.11539686.

Schouten, S. et al. (1995). "The effect of hydrosulphurization on stable carbon isotopic compositions of free and sulphur-bound lipids". In: *Geochimica et Cosmochimica Acta* 59.8, pp. 1605–1609. DOI: 10.1016/0016-7037(95)00066-9.

Schroder, S., J.P. Lacassie, and N.J. Beukes (2006). "Stratigraphic and geochemical framework of the Agouron drill cores, Transvaal Supergroup (Neoarchean-Paleoproterozoic, South Africa". In: *South African Journal of Geology* 109.1–2, pp. 23–54. DOI: 10.2113/gssajg.109.1-2.23.

Seewald, J.S., B.C. Benitez-Nelson, and J.K. Whelan (1998). *Laboratory and theoretical constraints on the generation and composition of natural gas.*

Sephton, M.A. (2002). "Organic compounds in carbonaceous meteorites". In: *Natural Product Reports* 19.3, pp. 292–311. DOI: 10.1039/b103775g.

Sherwood Lollar, B. et al. (2021). "A window into the abiotic carbon cycle – Acetate and formate in fracture waters in 2.7 billion year-old host rocks of the Canadian Shield". In: *Geochimica et Cosmochimica Acta* 294, pp. 295–314. DOI: 10.1016/j.gca.2020.11.026.

Shimoyama, A. and O. Matsubaya (1982). "Carbon Isotope Composition of Graphite and Carbonate in 3.8×10^9 Year Old Isua Rocks". In: *Chemistry Letters.*

Siahi, M. et al. (2016). "Sedimentology and facies analysis of Mesoarchaean stromatolitic carbonate rocks of the Pongola Supergroup, South Africa". In: *Precambrian Research* 278, pp. 244–264. DOI: 10.1016/j.precamres.2016.03.004.

Slotznick, S.P. and W.W. Fischer (2016). "Examining Archean methanotrophy". In: *Earth and Planetary Science Letters* 441, pp. 52–59. DOI: 10.1016/j.epsl.2016.02.013.

Snider, M.J. et al. (2002). "^{15}N kinetic isotope effects on uncatalyzed and enzymatic deamination of cytidine". In: *Biochemistry* 41.1, pp. 415–421. DOI: 10.1021/bi0114410i.

Soares, G.G. et al. (2019). "Phosphogenesis in the immediate aftermath of the Great Oxidation Event: Evidence from the Turee Creek Group, Western Australia". In: *Precambrian Research* 320, pp. 193–212. DOI: 10.1016/j.precamres.2018.10.017.

Strauss, H. and N.J. Beukes (1996). "Carbon and sulfur isotopic compositions of organic carbon and pyrite in sediments from the Transvaal Supergroup, South Africa". In: *Precambrian Research* 79.1–2, pp. 57–71. DOI: 10.1016/0301-9268(95)00088-7.

Stueken, E.E. et al. (2017). "Differential metamorphic effects on nitrogen isotopes in kerogen extracts and bulk rocks". In: *Geochimica et Cosmochimica Acta* 217, pp. 80–94. DOI: 10.1016/j.gca.2017.08.019.

Sugitani, K., K. Lepot, et al. (2010). "Biogenicity of Morphologically Diverse Carbonaceous Microstructures from the ca. 3400 Ma Strelley Pool Formation, in the Pilbara Craton, Western Australia". In: *Astrobiology* 10.9, pp. 899–920. DOI: 10.1089/ast.2010.0513.

Sugitani, K., K. Mimura, T. Nagaoka, et al. (2013). "Microfossil assemblage from the 3400Ma Strelley Pool Formation in the Pilbara Craton, Western Australia: Results form a new locality". In: *Precambrian Research* 226, pp. 59–74. DOI: 10.1016/j.precamres.2012.11.005.

Sugitani, K., K. Mimura, M. Takeuchi, et al. (2015). "A Paleoarchean coastal hydrothermal field inhabited by diverse microbial communities: the Strelley Pool Formation, Pilbara Craton, Western Australia". In: *Geobiology* 13.6, pp. 522–545. DOI: 10.1111/gbi.12150.

Sutherland, J.D. (2016). "The Origin of Life - Out of the Blue". In: *Angewandte Chemie - International Edition* 55.1, pp. 104–121. DOI: 10.1002/anie.201506585.

Tang, Y. et al. (2000). "Mathematical modeling of stable carbon isotope ratios in natural gases". In: *Geochimica et Cosmochimica Acta* 64 (15), pp. 2673–2687. DOI: 10.1016/S0016-7037(00)00377-X.

Tashiro, T. et al. (2017). "Early trace of life from 3.95 Ga sedimentary rocks in Labrador, Canada". In: *Nature* 549.7673, pp. 516–518. DOI: 10.1038/nature24019.

Thomazo, C. et al. (2009). "Methanotrophs regulated atmospheric sulfur isotope anomalies during the Mesoarchean (Tumbiana Formation, Western Australia". In: *Earth and Planetary Science Letters* 279.1–2, pp. 65–75. DOI: 10.1016/j.epsl.2008.12.036.

Tice, M.M. and D.R. Lowe (2004). "Photosynthetic microbial mats in the 3.416-Myr-old ocean". In: *Nature* 431.7008, pp. 549–552. DOI: 10.1038/nature02888.

Tissot, B.P. and D.H. Welte (1984). "Kerogen: composition and classification". In: *Petroleum formation and occurrence*, pp. 131–159.

Tocque, E. et al. (2005). "Carbon isotopic balance of kerogen pyrolysis effluents in a closed system". In: *Organic Geochemistry* 36.6, pp. 893–905. DOI: 10.1016/j.orggeochem.2005.01.007.

Ueno, Y. et al. (2002). "Ion microprobe analysis of graphite from ca. 3.8 Ga metasediments, Isua supracrustal belt, West Greenland: Relationship between metamorphism and carbon isotopic composition". In: *Geochimica et Cosmochimica Acta* 66.7, pp. 1257–1268. DOI: 10.1016/S0016-7037(01)00840-7.

Ungerer, P., J. Collell, and M. Yiannourakou (2015). "Molecular modeling of the volumetric and thermodynamic properties of kerogen: Influence of organic type and maturity". In: *Energy and Fuels* 29.1, pp. 91–105. DOI: 10.1021/ef502154k.

Valley, J.W. and J.R. O'Neil (1981). "$^{13}C^{12}C$ exchange between calcite and graphite: A possible thermometer in Grenville marbles". In: *Geochimica et Cosmochimica Acta* 45.3, pp. 411–419. DOI: 10.1016/0016-7037(81)90249-0.

Vandenbroucke, M. and C. Largeau (2007). "Kerogen origin, evolution and structure". In: *Organic Geochemistry* 38.5, pp. 719–833. DOI: 10.1016/j.orggeochem.2007.01.001.

Venkatachala, B.S. et al. (1990). "Archaean microbiota from the Donimalai formation, Dharwar supergroup, India". In: *Precambrian Research* 47.1–2, pp. 27–34. DOI: 10.1016/0301-9268(90)90028-0.

Vistoli, G. et al. (2013). "Advanced glycoxidation and lipoxidation end products (AGEs and ALEs): An overview of their mechanisms of formation". In: *Free Radical Research* 47.S1, pp. 3–27. DOI: 10.3109/10715762.2013.815348.

Wacey, D. (2010). "Stromatolites in the ~3400 Ma Strelley Pool Formation, Western Australia: Examining Biogenicity from the Macro- to the Nano-Scale". In: *Astrobiology* 10.4, pp. 381–395. DOI: 10.1089/ast.2009.0423.

Wacey, D., N. McLoughlin, et al. (2006). "The ~3.4 billion-year-old Strelley Pool Sandstone: a new window into early life on Earth". In: *International Journal of Astrobiology* 5.4, pp. 333–342. DOI: 10.1017/S1473550406003466.

Wacey, D., M. Saunders, and C. Kong (2018). "Remarkably preserved tephra from the 3430 Ma Strelley Pool Formation, Western Australia: Implications for the interpretation of Precambrian microfossils". In: *Earth and Planetary Science Letters* 487, pp. 33–43. DOI: 10.1016/j.epsl.2018.01.021.

Waldbauer, J.R. et al. (2009). "Late Archean molecular fossils from the Transvaal Supergroup record the antiquity of microbial diversity and aerobiosis". In: *Precambrian Research* 169.1–4, pp. 28–47. DOI: 10.1016/j.precamres.2008.10.011.

Walsh, M.M. (2004). "Evaluation of Early Archean Volcaniclastic and Volcanic Flow Rocks as Possible Sites for Carbonaceous Fossil Microbes". In: *Astrobiology* 4.4, pp. 429–437. DOI: 10.1089/ast.2004.4.429.

Walsh, M.M. and D.R. Lowe (1985). "Filamentous microfossils from the 3,500-Myr-old Onverwacht Group, Barberton Mountain Land, South Africa". In: *Nature* 314.6011, pp. 530–532. DOI: 10.1038/314530a0.

Walsh, M.W. (1992). "Microfossils and possible microfossils from the early archean onverwacht group, Barberton mountain land, South Africa". In: *Precambrian Research* 54.2–4, pp. 271–293. DOI: 10.1016/0301-9268(92)90074-X.

Walter, M.R., R. Buick, and J.S.R. Dunlop (1980). "Stromatolites 3,400–3,500 Myr old from the North Pole area, Western Australia". In: *Nature* 284.5755, pp. 443–445. DOI: 10.1038/284443a0.

Wang, R.Z. et al. (2023). "Carbon isotope fractionation by an ancestral rubisco suggests that biological proxies for CO_2 through geologic time should be reevaluated". In: *Proceedings of the National Academy of Sciences of the United States of America* 120.20. DOI: 10.1073/pnas.2300466120.

Ward, L.M., B. Rasmussen, and W.W. Fischer (2019). "Primary Productivity Was Limited by Electron Donors Prior to the Advent of Oxygenic Photosynthesis". In: *Journal of Geophysical Research: Biogeosciences* 124.2, pp. 211–226. DOI: 10.1029/2018JG004679.

Weiss, M.C. et al. (2016). "The physiology and habitat of the last universal common ancestor". In: *Nature Microbiology* 1.9, p. 16116. DOI: 10.1038/nmicrobiol.2016.116.

Westall, F. et al. (2001). "Early archean fossil bacteria and biofilms in hydrothermally-influenced sediments from the Barberton greenstone belt, South Africa". In: *Precambrian Research* 106.1–2, pp. 93–116. DOI: 10.1016/S0301-9268(00)00127-3.

Wiemann, J., J.M. Crawford, and D.E.G. Briggs (2020). "Phylogenetic and physiological signals in metazoan fossil biomolecules". In: *Science Advances* 6.28, pp. 1–11. DOI: 10.1126/sciadv.aba6883.

Wiemann, J., M. Fabbri, et al. (2018). "Fossilization transforms vertebrate hard tissue proteins into N-heterocyclic polymers". In: *Nature Communications* 9.1. DOI: 10.1038/s41467-018-07013-3.

Williford, K.H. et al. (2011). "Constraining atmospheric oxygen and seawater sulfate concentrations during Paleoproterozoic glaciation: In situ sulfur three-isotope microanalysis of pyrite from the Turee Creek Group, Western Australia". In: *Geochimica et Cosmochimica Acta* 75.19, pp. 5686–5705. DOI: 10.1016/j.gca.2011.07.010.

Wright, D.T. and W. Altermann (2000). "Microfacies development in Late Archaean stromatolites and oolites of the Ghaap Group of South Africa". In: *Geological Society, London, Special Publications* 178.1, pp. 51–70. DOI: 10.1144/GSL.SP.2000.178.01.05.

Wu, Z., R.P. Rodgers, and A.G. Marshall (2004). "Two- and Three-Dimensional van Krevelen Diagrams: A Graphical Analysis Complementary to the Kendrick Mass Plot for Sorting Elemental Compositions of Complex Organic Mixtures Based on Ultrahigh-Resolution Broadband Fourier Transform Ion Cyclotron Resonance Mass Measurements". In: *Analytical Chemistry* 76.9, pp. 2511–2516. DOI: 10.1021/ac0355449.

Xie, H., M. Formolo, and J. Eiler (2022). "Predicting isotopologue abundances in the products of organic catagenesis with a kinetic Monte-Carlo model". In: *Geochimica et Cosmochimica Acta* 327, pp. 200–228. DOI: 10.1016/j.gca.2022.03.028.

Yu, P.H. et al. (1981). "Deuterium isotope effects on the enzymatic oxidative deamination of trace amines". In: *Biochemical Pharmacology* 30.22, pp. 3089–3094.

Zeichner, S.S. (2023). szeichner/ArcheanKerogen. DOI: 10.5281/zenodo.8280728.

Zeichner, S.S., L.M Chimiak, et al. (2023). "Position-specific carbon isotopes of Murchison amino acids elucidate extraterrestrial abiotic organic synthesis networks". In: *Geochimica et Cosmochimica Acta*. DOI: 10.1016/j.gca.2023.06.010.

Zeichner, S.S., J. Ngheim, et al. (2021). "Early plant organics increased global terrestrial mud deposition through enhanced flocculation". In: *Science* 371 (6528), pp. 526–529. DOI: 10.1126/science.abd0379.

Zuilen, M.A. van et al. (2003). "Graphite and carbonates in the 3.8 Ga old Isua Supracrustal Belt, southern West Greenland". In: *Precambrian Research* 126.3–4, pp. 331–348. DOI: 10.1016/S0301-9268(03)00103-7.

Supplemental Methods: Maturation Model

The isotopic box model presented here offers a simplified interpretation of the chemical and resulting isotopic changes that will occur through diagenesis, catagenesis, and metagenesis, where the major changes in the isotopic composition of C within each process are driven by decarboxylation, homolytic cleavage, and exchange with dissolved inorganic carbon (DIC) in metamorphic fluids, respectively. The spreadsheet with all three model parts can be found in a supplementary spreadsheet (Zeichner, 2023).

The model requires interconversion between delta values, isotope ratio (R values), and fractional abundances (F values). We can interconvert between R and F using the following equation:

$$F = \frac{R}{(1 + R)} \tag{S5.1}$$

Diagenesis

We model the effect of decarboxylation on a model compound— decanoic acid—which we chose as a representation of the average O/C ratio in immature organic matter. While the relative contribution/preservation of sugars, lipids, and proteins deriving from biomass to deposited organic matter is still not fully understood, we chose relative amounts of 0.2, 0.6, and 0.2, respectively, informed by amounts reported in Burdige, 2007. These amounts not only result in an O/C ratio of 0.2, which is consistent with the starting point of immature OM in the Van Krevelin diagram, but also result in an average N/C ratio of 0.0667, which is consistent with the N/C ratio of immature organic matter reported by Wu, Rodgers, and Marshall, 2004.

It has been demonstrated in the literature that fatty acids like decanoic acid have intramolecular isotopic variation: The carboxyl group $\delta^{13}C$ value is higher than the $\delta^{13}C$ values of the other groups (-13‰ versus ~-27‰ for measurements of modern *E. coli*; Monson and Hayes, 1980). We therefore started out with the carboxyl site having $\delta^{13}C$ = -13‰ and the other 9 sites having $\delta^{13}C$ values of -37‰ (consistent with lower $\delta^{13}C$ values of Archean samples, as these values result a molecular average $\delta^{13}C$ of -34.6‰).

We iteratively model the decarboxylation of decanoic acid with model steps of 0.01 moles out of 1 total unreacted moles of fatty acid, where each step loses 2 oxygens for each C (as is the case with decarboxylation). We modeled a Rayleigh distillation process on the R value, which we converted from delta values using the following equation:

$$R = ((\delta/1000) + 1) + R_{VPDB} \qquad (S5.2)$$

where R_{VPDB}=0.01118. R values following the Rayleigh distillation process for each decarboxylation step were computed as follows:

$$R_{after} = R_{before} \times f_m^{(\alpha-1)} \qquad (S5.3)$$

where the f_m is the fraction of molecules unreacted, and α is the kinetic isotope effect (KIE) of decarboxylation. Here, we apply an α value = 0.97 ($^{13}k/^{12}k$) as reported in Lewis et al., 1993; Marlier and O'Leary, 1984. We convert the R_{after} for the site into an F value (Equation S5.1) and then use that F value to compute the average F value of the total organic carbons in the system, considering the instantaneous fraction of carboxyl sites relative to all carbons in the system (of both the decanoic acid and decarboxylated decanoic acid, i.e., nonane). The fraction of carboxyl sites (f_c) is calculated based on f_m as follows:

$$f_c = (f_m \times 1/10)/(1 - (1 - f_m) \times 1/10) \qquad (S5.4)$$

We assume that the average isotopic composition of neutral carbons remains constant during this reaction due to mass balance (although we note that this is a simplification, as decarboxylation reactions also isotopically discriminate against ^{13}C of the neutral carbon bonded to a carboxyl group). However, as the bonded neutral carbon does not leave the system, but is only transferred from the decanoic acid pool to the nonane pool, the average composition of neutral carbons in the system is preserved.

Finally, we compute an F value for the decanoic acid + nonane pool for each step, as follows:

$$F = F_{after} \times f_c + F_{alkane} \times (1 - f_c) \qquad (S5.5)$$

where F_{after} is the R_{after} value that was converted into an F value following Equation S5.1, and F_{alkane} is the fractional abundance of the original non carboxyl group carbon (i.e., $\delta^{13}C$ = -37‰). We can convert the F value into an R value, and then into its $\delta^{13}C$ value (assuming R_{VPDB}=0.01118 as noted above) to be able to understand how the isotope ratio would change over time within the context of a widely accepted international reference frame (VPDB).

We perform the above steps until the O/C ratio = 0.1, which is the approximate value where the O/C ratio of organic matter stops changing within the van Krevelin diagram. Our diagenesis model suggests that the bulk C isotopic composition of organic matter remains similar to its original composition, as a result of the competing consequences of the removal of a ^{13}C-rich site (i.e., carboxyl carbon), and the isotopic fractionation imparted by the decarboxylation reaction that causes ^{13}C enrichment in the remaining carboxyl carbons. Following diagenesis, catagenesis becomes the dominant phase of organic matter degradation, where elemental changes become mainly driven by the loss of H from the system.

Catagenesis

We use the final isotopic composition of our simplified organic matter from the diagenesis model as the starting point of the catagenesis model. The catagenesis model idealizes the maturation of C through the oil as being largely driven by homolytic cleavage, or the production of natural gas (Xie, Formolo, and Eiler, 2022).

We simplify our model using the following assumptions: We begin with a pool of 250 carbons, the number of carbons estimated to be within one unit cell of immature Type II kerogen (Ungerer, Collell, and Yiannourakou, 2015), and assume we have 300 H (for a starting H/C ratio = 1.2). We iteratively model the homolytic cleavage by assuming that each cracking reaction forms only methane, where 1 C and 4 H get removed at each reaction progress step (we note that in natural systems a mixture of methane, ethane, propane, and higher order hydrocarbons can be produced upon each cracking reaction; Xie, Formolo, and Eiler, 2022). For each step, we convert the initial $\delta^{13}C$ of the immature kerogen pool (we assume all the C have the same $\delta^{13}C$ value, equal to the final value of organic matter from the diagenesis model) to its R value. We assume that the reactant carbon pool undergoes Rayleigh distillation (following Equation S5.3), where in this case f_m is the fraction of unreacted carbons (out of the initial pool of 250 carbons), and α is the kinetic isotope effect (KIE) of homolytic cleavage. We apply an α value = 0.975 (^{13}k/^{12}k) as reported in Tang et al., 2000; Xie, Formolo, and Eiler, 2022. Finally, we convert the R value into its delta value as described above.

We iterate following the above steps until the H/C ratio of our organic matter pool = 0.1, similar to the H/C ratio of mature type II kerogen (Ungerer et al., 2015). Our homolytic cleavage model results in isotopic shifts of ~8‰. We also include data from Des Marais, 1997, which demonstrates a very similar result between our model and the prior empirical fit presented by Des Marais for the range of H/C values included in that paper.

Metamorphism

Finally, we model the effect of metamorphism on the isotopic composition of mature organic matter. This model is based on the observation that organic matter exposed to metamorphic conditions above greenschist facies can fully break down, exchange with reactive carbon in the metamorphic fluid, and recrystallize via a process of Ostwald ripening (Dunn and Valley, 1992; Kitchen and Valley, 1995). For simplicity, we assume that our organic matter starts with a $\delta^{13}C_{org}$ value equal to the final composition at the end of the catagenesis model, and exchanges with an infinitely large pool of dissolved inorganic carbon ($\delta^{13}C_{DIC}$ = 0‰), shifting the organic carbon towards an equilibrium exchange value of $\delta^{13}C_{org}$ = -6‰ (the expected value for organic matter that has been fully equilibrated with DIC at 600°C). As a result, this part of the model simply tracks the mixing of carbons with the original $\delta^{13}C_{org}$ value of organic matter, with the fully equilibrated graphitic end member ($\delta^{13}C_{org}$ = -6‰), while the reaction progress represents the fraction of carbon that underwent isotopic exchange with DIC.

For each reaction step, we convert the starting $\delta^{13}C$ values to R and then to F values (Equation S5.1), and compute the isotopic composition of the mixed product as follows:

$$F_{graphite} = f_{unreacted} \times F_{organic} + f_{reacted} \times F_{equil,graphite} \qquad \text{(S5.6)}$$

The computed $F_{graphite}$ values are then converted into R and $\delta^{13}C$ values. This exchange process drives large changes in $\delta^{13}C$ values of the organics, where at 100% reaction progress the $\delta^{13}C$ value of the organics = -6‰.

Supplemental Tables and Figures

Group or Formation	References
Towers Formation	Derenne et al., 2008; Schopf and Packer, 1987
Dresser Formation	Derenne et al., 2008; Morag et al., 2016
Apex Chert	Schopf and Packer, 1987
Panorama Group	Hofmann, Grey, et al., 1999
Strelley Pool Formation	Alleon et al., 2018; Allwood, Grotzinger, et al., 2009; Allwood, Kamber, et al., 2010; Allwood, Walter, Kamber, et al., 2006; Allwood, Walter, and Marshall, 2006; Duda et al., 2016; Flannery, Allwood, et al., 2018; Hofmann, Grey, et al., 1999; Lepot, Williford, et al., 2013; Lowe, 1983, 1994; Marshall et al., 2007; Sugitani, Lepot, et al., 2010; Sugitani, Mimura, Nagaoka, et al., 2013; Sugitani, Mimura, Takeuchi, et al., 2015; Wacey, 2010; Wacey, McLoughlin, et al., 2006; Wacey, Saunders, and Kong, 2018; Walter, Buick, and Dunlop, 1980
Onverwacht and Kromberg Fms	Hofmann and Bolhar, 2007; Kremer and Kaźmierczak, 2017; Muir and Hall, 1974; Tice and Lowe, 2004; Walsh, 2004; Walsh and Lowe, 1985; Walsh, 1992; Westall et al., 2001
Fig Tree Group	Byerly et al., 1996; Hofmann and Bolhar, 2007
Donimalai Formation	Venkatachala et al., 1990
Tumbiana Formation	Awramik and Buchheim, 2009; Bolhar and VanKranendonk, 2007; Coffey et al., 2013; Flannery and Walter, 2012; Lepot, Benzerara, Brown, et al., 2008; Lepot, Benzerara, Rividi, et al., 2009; Sakurai et al., 2005; Thomazo et al., 2009
Hamersley Group	Brocks, Buick, et al., 2003; Brocks, Summons, et al., 2003; Czaja, Johnson, et al., 2010; Ono et al., 2003; Partridge et al., 2008
Turee Creek	Barlow and Kranendonk, 2018; Fadel et al., 2017; Martindale et al., 2015; Murphy et al., 2016; Soares et al., 2019; Williford et al., 2011
Lime Acres Formation	Altermann and Schopf, 1995; Altermann and Wotherspoon, 1995
Gamohaan Formation	Altermann and Wotherspoon, 1995; Czaja, Beukes, and Osterhout, 2016; Fischer, Schroeder, et al., 2009; Gandin, Wright, and Melezhik, 2005; Johnson et al., 2003; Klein, Beukes, and Schopf, 1987; Rivera and Sumner, 2014
Pongola Supergroup	Beukes and Lowe, 1989; Siahi et al., 2016
Moodies Group	Bontognali, Fischer, and Föllmi, 2013; Heubeck, 2009; Heubeck and Lowe, 1994; Homann, Heubeck, Airo, et al., 2015; Homann, Heubeck, Bontognali, et al., 2016; Noffke, Eriksson, et al., 2006
Mozaan Group	Noffke, Hazen, and Nhleko, 2003; Ono, 2006
Papers reviewing microfossils of acritarchs and algae post-GOE	Cohen and Macdonald, 2015; Javaux, Knoll, and Walter, 2004; Knoll et al., 2006; Porter, 2004

Table S5.1: References of studies examining sedimentology and fossiliferous textures within Archean units.

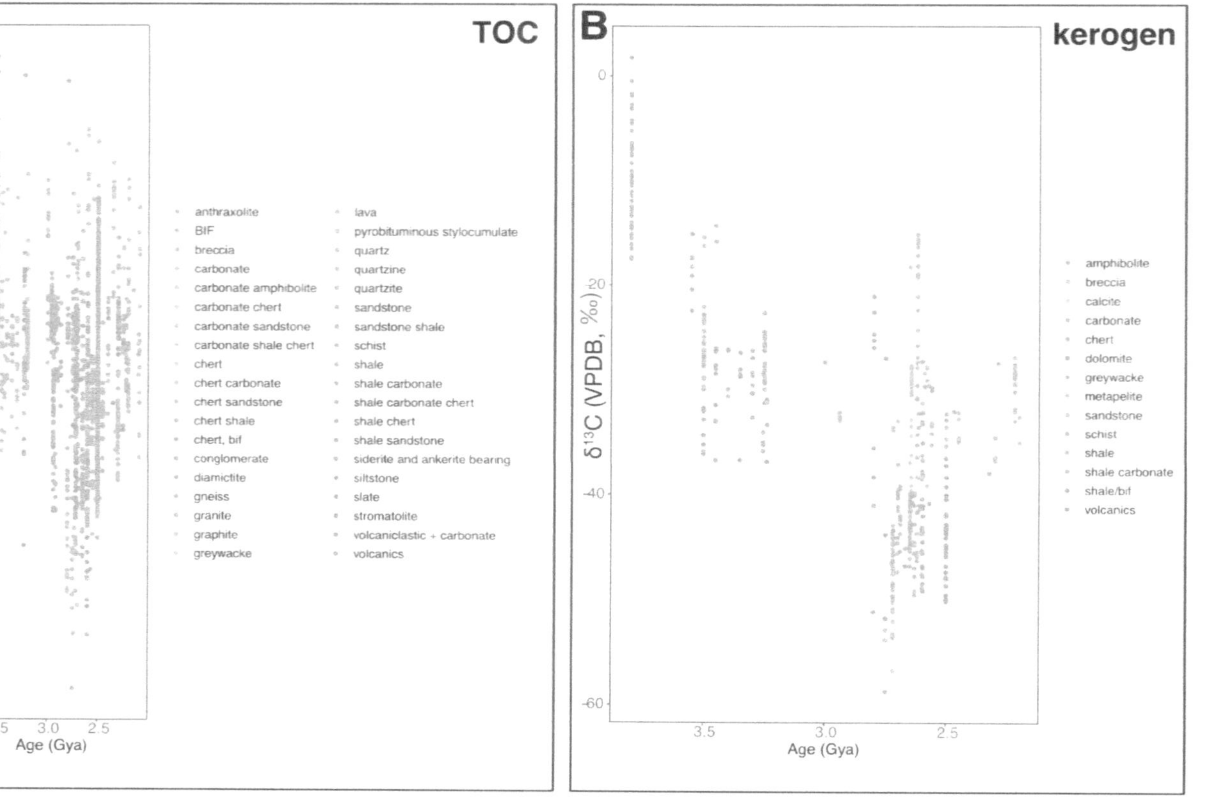

Fig. S5.1: Archean kerogen carbon isotopes by dominant lithology.

Compiled $\delta^{13}C_{VPBD}$ values for (A) total organic carbon and (B) kerogen measured from rocks prior to the GOE, color coded by dominant lithology. References are included in the supplementary material. TOC; total organic carbon.

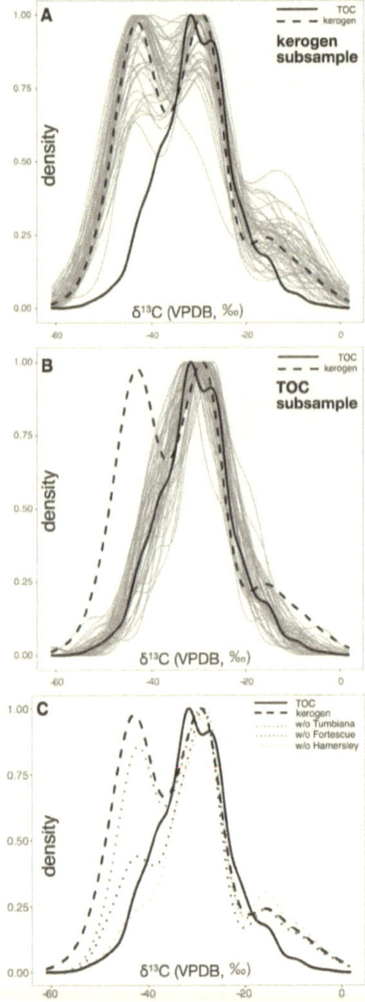

Fig. S5.2: Subsampling $\delta^{13}C$ values of kerogen and TOC.

50 iterations of 100 point subsamples of $\delta^{13}C$ values of (A) kerogen and (B) TOC, compared with all TOC data (black line) and kerogen data (black dashed line). (C) Comparison of kernel density estimates for $\delta^{13}C$ values of TOC and kerogen with those of kerogen data removing Tumbiana Fm. measurements (red dotted line), Fortescue Group measurements (blue dotted line), and Hamerlsey measurements (orange dotted line). TOC; total organic carbon, w/o; without.

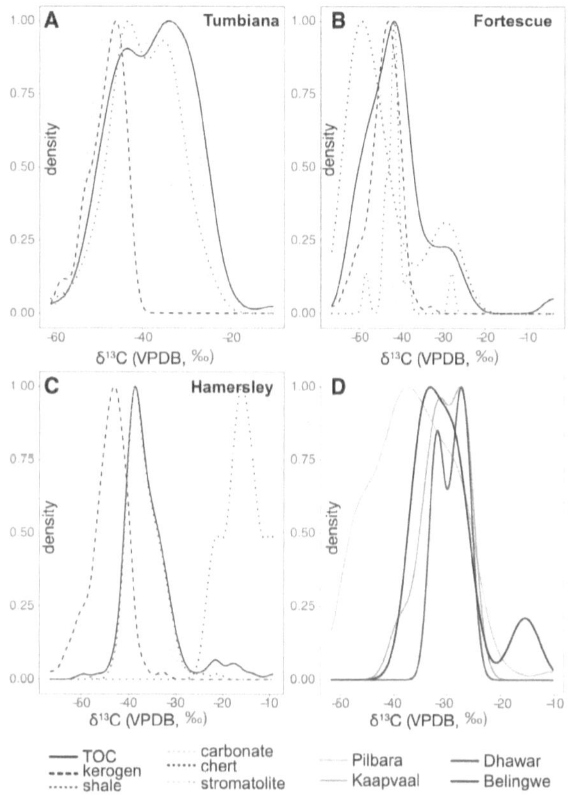

Fig. S5.3: Case study of 2.7 Gya organic carbon.

Kernel density estimates for $\delta^{13}C$ values of kerogen (dashed black line) and TOC (black line) versus lithology-specific $\delta^{13}C$ values of TOC (dotted lines) for (A) Tumbiana Fm., (B) Fortescue Group, (C) Hamersley Group. (D) Comparison of the kernel density estimate for $\delta^{13}C$ values of TOC for units from different cratons with samples 2.65-2.8Gya. TOC; total organic carbon.

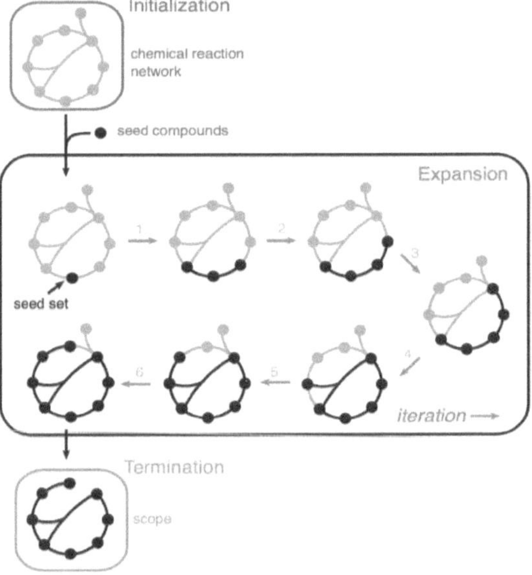

Fig. S5.4: The network expansion algorithm.

The network expansion algorithm (Ebenhöh, Handorf, and Heinrich, 2004; Handorf, Ebenhöh, and Heinrich, 2005) produces networks of metabolites (nodes) and edges (reactions) from a set of "seed" compounds and an initial chemical reaction network (gray network, top left). The set of compounds in the seed set are used to determine whether each reaction in the network is feasible, which depends on both substrate and cofactor availability. Reactions that are feasible are allowed to generate product compounds at each expansion iteration (green lines). These product compounds are added to the seed set, and the process is repeated. The algorithm terminates when no additional compounds can be added to the network, producing a subset of compounds and reactions (black) from the original (gray) network. The set of compounds and reactions at the end of the expansion process is called the "scope".

Chapter 6

EARLY PLANT ORGANICS INCREASED GLOBAL TERRESTRIAL MUD DEPOSITION THROUGH ENHANCED FLOCCULATION

S.S. Zeichner, J. Ngheim, et al. (2021). "Early plant organics increased global terrestrial mud deposition through enhanced flocculation". In: *Science* 371 (6528), pp. 526–529. DOI: `10.1126/science.abd0379`.

Sarah Zeichner[1], Justin Nghiem[1], Michael P. Lamb[1], Nina Takashima[1], Jan de Leeuw[1], Vamsi Ganti[2,3], Woodward W. Fischer[1]

Affiliations: [1] Division of Geological & Planetary Sciences, California Institute of Technology, Pasadena, CA [2] Department of Geography, University of California Santa Barbara, Santa Barbara, CA [3] Department of Earth Science, University of California Santa Barbara, Santa Barbara, CA

The Paleozoic evolution and proliferation of terrestrial plants has been connected with changes in soil and atmospheric chemistry, and increased land primary productivity and organic carbon deposition (Boyce and Lee, 2017). Correspondingly, the stratigraphic record contains a major first-order change in the construction of river floodplain deposits (Long, 1977). A recent study quantified an Ordovician-Silurian increase in alluvial mudrock , that is, siliciclastic rock consisting of at least 50% mud-sized particles, which occurred concurrently with the evolution of early plants (Fig. 6.1A; McMahon and Davies, 2018). However, these data presented an apparent paradox. One explanation could be that the proliferation of plants led to mud production, yet the sedimentary record contains abundant mudrock throughout Earth history, albeit in marine paleoenvironments prior to early Paleozoic time (Davies and Gibling, 2010). Further, this increase in alluvial mudrock predated the evolution of large rooted plants and forests (Fig. 6.1A; Fischer, 2018; McMahon and Davies, 2018. Early plants were small in size (~ 1 cm tall) and lacked deep rooting (Boyce and Lee, 2017) that was likely necessary for floodplain binding through rooting and flow baffling (Hadley, 1961).

We hypothesized that these plants instead could have increased mudrock via a molecular mechanism: the rise in terrestrial organic material associated with early plants would drive mud flocculation in rivers and, in consequence, enhance mud settling and deposition on river floodplains. Plant polymers are not uniquely capable of binding sediment—previous work demonstrated the presence of pre-Silurian terrestrial microbiota building sedimentary structures (McMahon and Davies, 2018). However, the proliferation of early plants dramatically increased productivity of the land surface, and thereby the flux of organic polymers in terrestrial environments—both the polymers produced directly by the plants themselves, as well as those generated and further modified by the rich associated microbial communities (Boyce and Lee, 2017; Fischer, 2018).

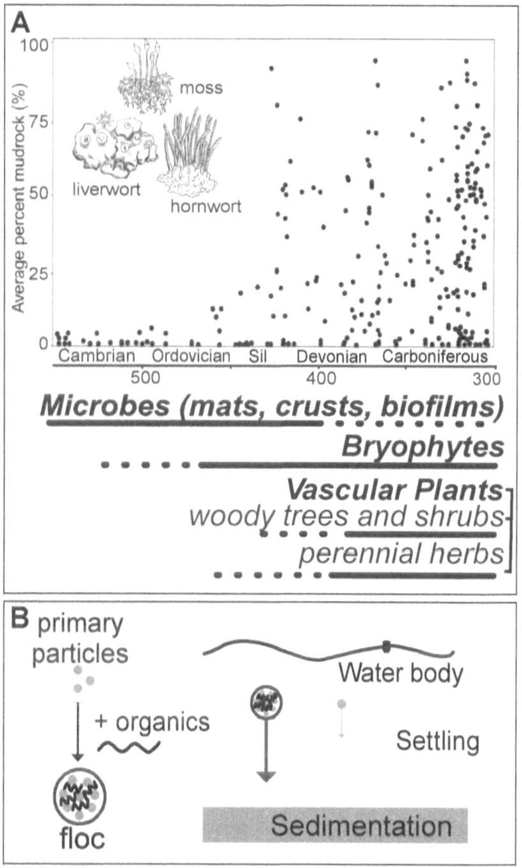

Figure 6.1: Mudrock abundance, plant evolution, and flocculation.

(A) Average percent mudrock for alluvial deposits over time, adapted from McMahon and Davies, 2018, shown alongside ranges of key evolutionary events in early plant evolution (Boyce and Lee, 2017). The amount of alluvial mudrock increased with proliferation of early diverging plant lineages (e.g. bryophytes: liverworts, hornworts, and mosses). (B) Organic polymers (black squiggles) can bind primary clay particles (gray dots) together into flocs (black circles), which increases their settling velocities and sedimentation rates. The size of the thick, black arrow indicates increased settling velocities driven by larger particle aggregate surface areas upon flocculation.

Flocculation is the process of binding of individual particles into larger aggregates called "flocs" (Fig. 6.1B), and is known to promote the deposition of clay and silt (i.e., mud) within estuarine and marine environments (Maggi, 2005). Flocculation can substantially increase mud settling velocities (Zhang et al., 2013), and growing evidence points to mud flocculation occurring in modern rivers (Fig. 6.1B; Lamb, Leeuw, et al., 2020). Therefore, an increase in the ability to flocculate fluvial sediment has the potential to cause a major rise in alluvial mud deposition rates.

Flocculation in freshwater is associated with the presence of organics (Zhang et al., 2013) because they—particularly polymers—facilitate particle binding interactions (Sholkovitz, 1976). Laboratory studies found that combinations of primary particles and polymers could lead to variable levels of flocculation (Furukawa, Reed, and Zhang, 2014; Zhang et al., 2013). However, these studies did not directly measure the effect of polymer-clay combinations on floc settling velocity. Evidence for widespread flocculation in natural rivers came from an analysis of suspended sediment concentration-depth profiles, which showed systematically larger settling velocities compared to theoretical expectations for sediment smaller than 40 μm in diameter (Lamb, Leeuw, et al., 2020). Thus, flocculation appears to be a primary control on mud settling in modern fluvial environments, but the specific roles of organic matter in driving flocculation in rivers have remained unclear.

We conducted 83 flume experiments with distinct combinations of model organic polymers (xanthan gum and guar gum) and clay minerals (smectite and kaolinite) (Fig. 6.2A inset) to quantify the role of organic material in determining mud settling rates in freshwater rivers (Furukawa, Reed, and Zhang, 2014). The experiments had constant turbulent Reynolds numbers , volumetric sediment concentrations of ~ 0.1 g L^{-1} and low ionic strength, similar to natural rivers (Table S6.1; Supplementary materials and methods). Although these abiotic variables can also affect flocculation, particularly in marine and estuarine settings (Furukawa, Reed, and Zhang, 2014), our goal was to isolate the effect of organics on freshwater flocculation. We performed experiments in a fixed volume stirred batch reactor within a light-sensitive box. For each experiment, we mixed specific proportions of clay and organic polymer, fully suspended the sediment in the water (Fig. 6.2A inset), and then captured time-lapse photographs of suspended sediment once the turbulent mixing was stopped. We calibrated the time-series absorbance data to derive sediment concentration, and regressed the concentration data on time to calculate settling velocities (Fig. 6.2A).

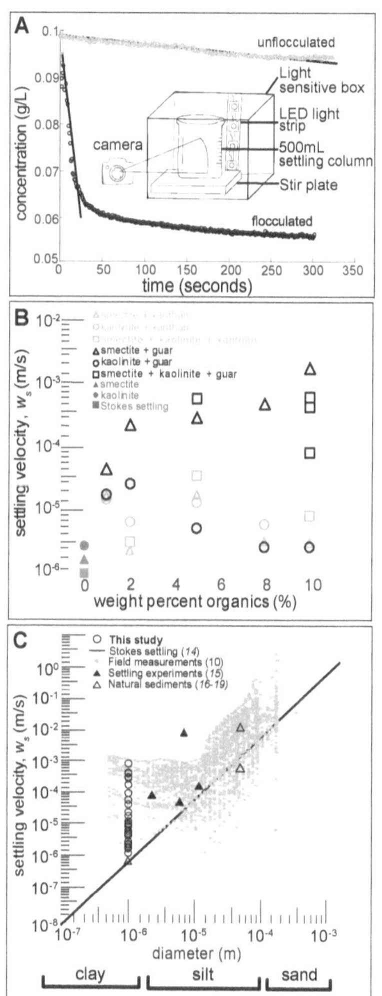

Figure 6.2: Experimental results.

Caption continued on next page.

Figure 6.2: Experimental results.

(A) Suspended sediment concentration as a function of time from an example experiment with smectite control experiment (gray) compared to an experiment of 5 wt.% guar gum with smectite (black). Floc settling velocity was determined from the rate of concentration change in time, as shown by the fit black line (Supplementary Materials and Methods). Inset shows the experiment setup. (B) Settling velocities measured from our experiments as a function of weight percentage organics for different combinations of clays and organics, compared to the Stokes' settling rate for primary (unflocculated) particles (Stokes, 1851). (C) Settling velocities from our experiments and previous work as a function of primary particle diameter, and the Stokes' prediction for unflocculated particles (Stokes, 1851). Previous studies include settling experiments with clay and organic polymers comparable to our experiments (Tan et al., 2012), experiments with natural sediments (Lick, Huang, and Jepsen, 1993; Námer and Ganczarczyk, 1993; Shang et al., 2014; Wendling et al., 2015), and settling velocities from modern rivers inverted from concentration-depth profiles (Lamb, Leeuw, et al., 2020).

We observed that organic polymers had a substantial, varied, and non-linear effect on clay flocculation and settling velocity. All experiments with organic polymers formed visible flocs (Fig. S6.1), which settled significantly faster than primary unflocculated clay particles (primary particle median diameter $D_{50} = 1$ μm; settling velocity, $w_{s,\text{unflocculated}} = 2.205 \times 10^{-6}$ m/s; p-value = 0.004) (Fig. 6.2B; Supplementary Materials and Methods). Generally, floc settling velocities increased with organic concentrations (Table S6.2; ; Supplementary Materials and Methods). Guar gum was a more effective flocculant compared to xanthan gum (p-value = 0.004), likely due to its charge and branched structure that increased the number of possible crosslinks between clays and organics. Together, guar gum and smectite formed the largest flocs with the fastest settling velocities ($w_s \sim 10^{-3}$ m/s; p-value = 0.001; Fig. 6.2B), even when mixed with kaolinite. Additional particle tracking experiments using humic acids and kaolinite also yielded readily observable flocs and settling velocities of $\sim 1 \times 10^{-3}$ m/s—albeit at much higher sediment concentrations (Supplementary Materials and Methods). Settling velocities from our floc experiments deviated substantially from Stokes' settling velocity predictions (Stokes, 1851) for clay primary particles by up to three orders of magnitude (Fig. 6.2B,C), and instead yielded velocities expected for medium to coarse silt (D = 20-63 μm).

Our experimental results were consistent with data from previous experiments that characterized freshwater flocculation for similar combinations of clay with organic polymers and demonstrated both qualitatively and quantitatively (Fig. 6.2C; Tan et al., 2012; Zhang et al., 2013) optimal conditions for flocculation (Fig. 6.2B). The results also agreed with previous measurements of freshwater floc settling velocities from activated sludge (Námer and Ganczarczyk, 1993) and natural sediment from rivers (Fig. 6.2C; Lick, Huang, and Jepsen, 1993; Námer and Ganczarczyk, 1993; Shang et al., 2014; Wendling et al., 2015). In particular, the guar-smectite experiments produced settling velocities consistent with those estimated in rivers by inversion of suspended sediment concentration-depth profiles (Fig. 6.2B, C; Lamb, Leeuw, et al., 2020).

The mineralogical and organic materials found in rivers can be more complex than we simulated in our experiments. However, we found that interactions between clays with charged interlayers and charged branched polymers can produce settling velocities like those observed in rivers ($w_s \sim 10^{-4}$ to 10^{-3} m/s) (Lamb, Leeuw, et al., 2020), even if we added other clays like kaolinite to the mixture. Experiments without organics or with low organic concentrations did not produce enhanced settling velocities (Fig. S6.1); these conditions are atypical for modern rivers. Likewise, the presence of silt common to rivers would further increase floc settling velocities compared to those measured in our experiments (Tran and Strom, 2017). Although we used idealized, chemically well-defined polymers, these polymers have comparable structures and functional groups to a range of plant-derived materials (Supplementary Materials and Methods), including those found in modern plant cell walls (Pauly and Keegstra, 2010), and are thought to have remained relatively consistent throughout plant evolution (Popper, 2008). This similarity supported the notion that organics play an important role in mud sediment transport in rivers, and vice versa.

We used a 1-D advection-settling analytical model to study the effect of organic-driven flocculation on overbank floodplain deposition, scaled roughly after the Mississippi River (Fig. 6.3A; Supplementary Materials and Methods) as an example of the deep channeled, low-gradient single-threaded rivers common before and after the Silurian Period (Ganti, Whittaker, et al., 2019). Model results showed systematically higher mud abundance relative to that of sand across the floodplain width in a flocculated scenario compared to an unflocculated scenario (26% mud for flocculated compared to 15% for unflocculated near the channel; Fig. 6.3C). For the flocculated case, we assumed that particles with D < 20 µm settled at a rate of 0.34 mm/s (Fig. 6.3B), similar to observations from rivers (Lamb, Leeuw, et al.,

2020, Supplementary Materials and Methods). Mud abundances in the flocculated case exceeded 50%—the definition of mudrock—everywhere beyond ~ 80 m of the channel edge. Both scenarios predicted predominantly mud deposition beyond ~ 200 m of the channel because this is beyond the advection-settling length for sand in our model (Fig. 6.3C; Supplementary Materials and Methods). Importantly, flocculation also caused two-fold higher mud deposition rates at kilometer-scale distances from the channel (Fig. 6.3D). These model results can be generalized to any river system by changing the overbank water discharge, sediment concentration, and the sediment size distribution (Fig. S6.4; Supplementary Materials and Methods). While changing parameter values affected the mud deposition rate and the transition location from sandstone to mudrock, the general result of flocculation resulting in muddier overbank deposition held for all scenarios (Supplementary Materials and Methods).

Our results have substantial implications for the proliferation of fluvial mudrock. The model predicted that flocculation results in muddier channel banks (Fig. 6.3C), which can increase bank cohesion and reduce the channel lateral migration rates (Lapôtre et al., 2019). Slower lateral migration rates, in turn, limit the width of sandy channel-belt deposits (Jerolmack and Mohrig, 2007). Furthermore, muddier channel-proximal deposits can cause channel narrowing, restrict braiding, and decrease channel sinuosity (Lapôtre et al., 2019), all of which limit the extent of sandy channel-belt deposits. When this deposition pattern is spatially superposed over time as channels aggrade and migrate laterally, these feedbacks should produce overall muddier floodplains than predicted by our simple model that lacked channel dynamics. At the larger basin scale, increased rates of mud deposition kilometers from the channel due to flocculation (Fig. 6.3D) will make mud preservation overall more likely, with lower chances of reworking by fluvial or aeolian processes (Fig. S6.3; Supplementary Materials and Methods; Ganti, Hajek, et al., 2020). In addition, the rates of lateral migration relative to channel switching, or avulsion, determine the stacking pattern and preservation of channel-belt sandstone bodies (Jerolmack and Mohrig, 2007). Thus, slower lateral migration rates not only reduce the extent of individual sandstone bodies, but also shift the alluvial architecture to a pattern characterized by mudrock with isolated sandstone bodies (Fig. S6.3; Supplementary Materials and Methods, Jerolmack and Mohrig, 2007). In contrast, the architecture of Precambrian alluvial deposits is characterized by laterally extensive sandstone bodies with substantial amalgamation and low mudrock preservation (Cotter, 1978;

Ielpi et al., 2018; McMahon and Davies, 2018; Winston, 1978)—features indicative of high rates of channel lateral migration, relative to channel avulsion (Ganti, Hajek, et al., 2020).

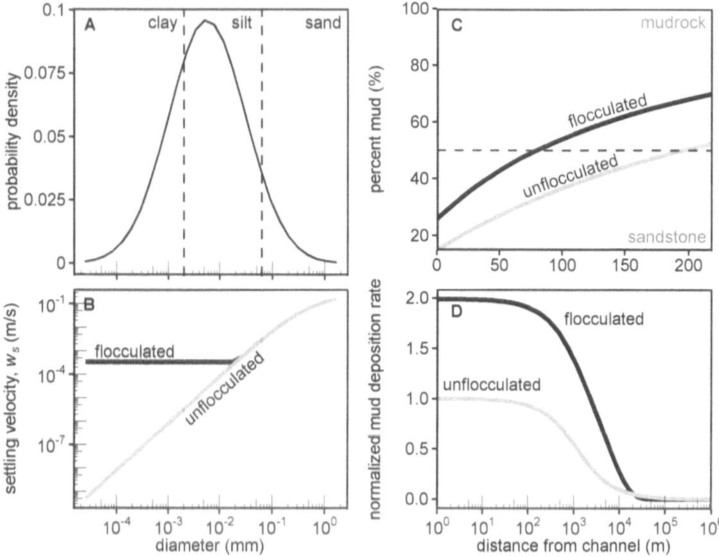

Figure 6.3: Floodplain sediment transport model.

(A) Grain size distribution of sediment supplied from the channel onto the floodplain, approximately scaled after the Mississippi River (Supplementary Materials and Methods). (B) Settling velocities used in the model for flocculated and unflocculated scenarios. (C) Results of percent mud in the proximal floodplain, classified as mudrock and sandstone. (D) Mud deposition rate as a function of distance from channel, normalized by the rate at the channel edge for the unflocculated case.

Our experiments illustrated how plant-associated organics can cause flocculation and substantially increase the settling velocity of mud in freshwater rivers, resulting in muddier floodplains. This coupling between organic carbon and mud transport and deposition has important implications for carbon cycling and sequestration. The proliferation of early land plants in terrestrial ecosystems during the Ordovician and Silurian periods increased the amount of primary production and the burial flux of organic material by at least an order of magnitude (Edwards, Cherns, and Raven,

2015). This increase in the amount of organic matter in terrestrial environments would have generated a diverse suite of polymeric molecules, both via direct synthesis from plants as well as the microbial communities that thrive on plant-derived organic matter (Cheshire, 1977). Together, all these polysaccharides would drive the binding of fine sediment particles. Thus, a natural correlate of a plant-driven flocculation mechanism would have been an increase in organic carbon content in fluvial deposits associated with the increase in mudrock in those deposits. Although only few examples are suitable for comparison, and total organic carbon (TOC) might be lower on average in older rocks due to preservation biases, existing geochemical data supported the idea that pre-Ordovician rivers had lower TOC contents and thus relatively less effective mud flocculation (and sandier floodplains). Proterozoic alluvial rocks with low mudrock abundances have commensurately low TOC (e.g. Nonesuch Formation: <1% alluvial mudrock and <1% TOC; Hatch and Morey, 1985). In contrast, alluvial rocks post-dating the evolution of land plants have both greater mudrock abundances and greater TOC concentrations within those mudrocks (e.g., lower Carboniferous Hørybreen/Mormien Formation; up to 28% mudrock and 11-30% TOC; Dallmann et al., 2004; McMahon and Davies, 2018). These trends are consistent with the hypothesis that plant proliferation, with its commensurate rise in primary production and terrestrial organic carbon fluxes, could have contributed substantially to the abrupt and irreversible early Paleozoic increase in alluvial mudrocks (McMahon and Davies, 2018). Likewise, the processes of flocculation and enhanced settling of mud onto floodplains outlined an efficacious mechanism for organic carbon burial in ancient and modern alluvial systems.

Funding: This work was made possible with support from the Caltech Discovery Fund (to WF and ML), David and Lucile Packard Foundation (WF), American Chemical Society Petroleum Research Fund (WF), National Science Foundation Graduate Research Fellowship Program (SZ), and Troy Tech High School Program (NT).

Author Contributions: Sarah Zeichner: Methodology, Formal analysis, Software, Writing, Visualization. Justin Nghiem: Software, Formal analysis, Writing, Visualization. Michael Lamb: Conceptualization, Supervision, Funding acquisition. Jan de Leeuw: Methodology, Writing (Review Editing). Nina Takashima: Validation, Data curation. Vamsi Ganti: Formal analysis, Writing (Review Editing). Woodward Fischer: Conceptualization, Supervision, Funding acquisition.

Data and material availability: All data is available in the main text or the supplementary materials. All code for image processing, statistical analyses and 1-D sedimentation model is available online (Zeichner, Nghiem, et al., 2020).

References

Boyce, C.K. and J.-E. Lee (2017). "Plant Evolution and Climate Over Geological Timescales". In: *Annual Reviews in Earth and Planetary Sciences* 45, pp. 61–87. DOI: 10.1146/annurev-earth-063016.

Cheshire, M.V. (1977). "Origins and Stability of Soil Polysaccharide". In: *Journal of Soil Science* 28, pp. 1–10. DOI: 10.1111/j.1365-2389.1977.tb02290.x.

Cotter, E. (1978). *Fluvial Sedimentology*. Ed. by A.D. Miall.

Dallmann, E. et al. (2004). "Lithostratigraphic lexicon of Svalbard: Review and recommendations for nomenclature use: Upper Palaeozoic to Quaternary bedrock Upper Palaeozoic lithostratigraphy". In: *Committee on the Stratigraphy of Svalbard* 166, pp. 29–65.

Davies, N.S. and M.R. Gibling (2010). "Cambrian to Devonian evolution of alluvial systems: The sedimentological impact of the earliest land plants". In: *Earth Science Reviews* 98, pp. 171–200. DOI: 10.1016/j.earscirev.2009.11.002.

Edwards, D., L. Cherns, and J.A. Raven (2015). "Could land-based early photosynthesizing ecosystems have bioengineered the planet in mid-Palaeozoic times?" In: *Palaeontology* 58, pp. 803–837. DOI: 10.1111/pala.12187.

Ferguson, R.I. and M. Church (2007). "A Simple Universal Equation for Grain Settling Velocity". In: *Journal of Sedimentary Research* 74, pp. 933–937. DOI: 10.1306/051204740933.

Fischer, W.W. (2018). "Early plants and the rise of mud". In: *Science* 359, pp. 994–995. DOI: 10.1126/science.aas9886.

Furukawa, Y., A.H. Reed, and G. Zhang (2014). "Effect of organic matter on estuarine flocculation: A laboratory study using montmorillonite, humic acid, xanthan gum, guar gum and natural estuarine flocs". In: *Geochemical Transactions* 15, pp. 1–9. DOI: 10.1186/1467-4866-15-1.

Galy, V., B. Peucker-Ehrenbrink, and T. Eglinton (2015). "Global carbon export from the terrestrial biosphere controlled by erosion". In: *Nature* 521, pp. 204–207. DOI: 10.1038/nature14400.

Ganti, V., E.A. Hajek, et al. (2020). "Morphodynamic Hierarchy and the Fabric of the Sedimentary Record". In: *Geophysical Research Letters* 47, pp. 1–10. DOI: 10.1029/2020GL087921.

Ganti, V., A.C. Whittaker, et al. (2019). "Low-gradient, single-threaded rivers prior to greening of the continents". In: *Proceedings of the National Academy of Sciences of the United States of America*. Vol. 116, pp. 11652–11657. DOI: 10.1073/pnas.1901642116.

Hadley, R.F. (1961). "Influence of riparian vegetation on channel shape, northeastern Arizona". In: *US Geological Survery Professional Paper* 424–C, pp. 30–31.

Hatch, J.R. and G.B. Morey (1985). "Hydrocarbon Source Rock Evaluation of Middle Proterozoic Solor Church Formation, North American Mid-Continent Rift System, Rice County, Minnesota". In: *American Association of Petroleum Geologists Bulletin* 69, pp. 1208–1216. DOI: `10.1016/j.cossms.2010.07.001`.

Ielpi, A. et al. (2018). "Fluvial floodplains prior to greening of the continents: Stratigraphic record, geodynamic setting, and modern analogues". In: *Sedimentary Geology* 372, pp. 140–172. DOI: `10.1016/j.sedgeo.2018.05.009`.

Ito, A. and R. Wagai (2017). "Data Descriptor: Global distribution of clay-size minerals on land surface for biogeochemical and climatological studies". In: *Scientific Data* 4, pp. 1–11. DOI: `10.1038/sdata.2017.1031`.

Jerolmack, D.J. and D. Mohrig (2007). "Conditions for branching in depositional rives". In: *Geology* 35, pp. 463–466. DOI: `10.1130/G23308A.1`.

Jordan, B.P.R. (1965). "Fluvial Sediment of the Mississippi River at St Louis, Missouri". In: *United States Government Printing Office*, pp. 1–98.

King, R., ed. (1992). *Fluid Mechanics of Mixing: Modelling, Operations and Experimental Techniques*. Vol. 10. DOI: `10.1007/978-94-015-7973-5_3`.

Kolmogorov, A.N. (1991). "The Local Structure of Turbulence in Incompressible Viscous Fluid for Very Large Reynolds Numbers". In: *Proceedings: Mathematical and Physical Sciences* 434.

Lamb, M.P., J.De Leeuw, et al. (2020). "Mud is transported as flocculated and suspended bed material". In: *Nature Geoscience* 13, pp. 566–570.

Lamb, M.P., B. McElroy, et al. (2010). "Linking river-flood dynamics to hyperpycnal-plume deposits: Experiments, theory, and geological implications". In: *Bulletin of the Geological Society of America* 122, pp. 1389–1400. DOI: `10.1130/B30125.1`.

Lapôtre, M.G.A. et al. (2019). "Model for the Formation of Single-Thread Rivers in Barren Landscapes and Implications for Pre-Silurian and Martian Fluvial Deposits". In: *Journal of Geophysical Research: Earth Surface* 124, pp. 2757–2777. DOI: `10.1029/2019JF005156`.

Le Quéré, C.B. et al. (2018). "Global Carbon Budget 2018". In: *Global Carbon Project* 4, pp. 2141–2194. DOI: `10.18160/gcp-2018`.

Lick, W., H. Huang, and R. Jepsen (1993). "Flocculation of fine-grained sediments due to differential settling". In: *Journal of Geophysical Research* 98, p. 10279. DOI: `10.1029/93jc00519`.

Long, D. (1977). "Proterozoic Stream Deposits: Some Problems of Recognition and Interpretation of Ancient Sandy Fluvial Systems". In: *Fluvial Sedimentology Memoir* 5, pp. 313–341.

Maggi, F. (2005). "Flocculation dynamics of cohesive sediment". PhD book. Faculty of Civil Engineering and Geosciences, Delft University of Technology.

McMahon, W.J. and N.S. Davies (2018). "Evolution of alluvial mudrock forced by early land plants". In: *Science* 359, pp. 1022–1024. DOI: 10.1126/science.aan4660.

Námer, Juraj and Jerzy J. Ganczarczyk (1993). "Settling properties of digested sludge particle aggregates". In: *Water Research* 27 (8), pp. 1285–1294. DOI: 10.1016/0043-1354(93)90215-4.

Paola, C. and V.R. Voller (2005). "A generalized Exner equation for sediment mass balance". In: *Journal of Geophysical Research: Earth Surface* 110, pp. 1–8. DOI: 10.1029/2004JF000274.

Parker, G. (1978). "Self-formed straight rivers with equilibrium banks and mobile bed. Part 1. The sand-silt river". In: *Journal of Fluid Mechanics* 89, pp. 109–125. DOI: 10.1017/S0022112078002499.

Parker, G. et al. (1987). "Experiments on turbidity currents over an erodible bed". In: *Journal of Hydraulic Research* 25, pp. 123–147. DOI: 10.1080/00221688709499292.

Pauly, M. and K. Keegstra (2010). "Plant cell wall polymers as precursors for biofuels". In: *Current Opinion in Plant Biology* 13, pp. 304–311. DOI: 10.1016/j.pbi.2009.12.009.

Popper, Z.A. (2008). "Evolution and diversity of green plant cell walls". In: *Current Opinion in Plant Biology* 11, pp. 286–292. DOI: 10.1016/j.pbi.2008.02.012.

Scott, C.H. and H.D. Stephens (1966). "Special sediment investigations: Mississippi River at St Louis, Missouri, 1961-63". In: *US Geological Survey*.

Shang, Q.Q. et al. (2014). "Biofilm effects on size gradation, drag coefficient and settling velocity of sediment particles". In: *International Journal of Sediment Research* 29, pp. 471–480. DOI: 10.1016/S1001-6279(14)60060-3.

Sholkovitz, E.R. (1976). "Flocculation of dissolved organic and inorganic matter during the mixing of river water and seawater". In: *Geochimica et Cosmochimica Acta* 40, pp. 831–845. DOI: 10.1016/0016-7037(76)90035-1.

Stokes, G.G. (1851). "On the effect of internal friction of fluids on the motion of pendulums". In: *Transactions of the Cambridge Philosophical Society* 9, pp. 8–106.

Tan, X.L. et al. (2012). "Characterization of particle size and settling velocity of cohesive sediments affected by a neutral exopolymer". In: *International Journal of Sediment Research* 27, pp. 473–485. DOI: 10.1016/S1001-6279(13)60006-2.

Tran, D. and K. Strom (2017). "Suspended clays and silts: Are they independent or dependent fractions when it comes to settling in a turbulent suspension?" In: *Continental Shelf Research* 138.

Wang, C.F. et al. (2017). "Characterization of humic acids extracted from a lignite and interpretation for the mass spectra". In: *RSC Advances* 7, pp. 20677–20684. DOI: 10.1039/C7RA01497J.

Wendling, V. et al. (2015). "Using an optical settling column to assess suspension characteristics within the free, flocculation, and hindered settling regimes". In: *Journal of Soils and Sediments* 15, pp. 1991–2003. DOI: 10.1007/s11368-015-1135-1.

Wershaw, R.L. (2004). "Evaluation of conceptual models of natural organic matter (humus) from a consideration of the chemical and biochemical processes of humification". In: *USGS Report* 5121.

Winston, D. (1978). "Fluvial Sedimentology". In: *Canadian Society of Petroleum Geologists Memoir 5*. Ed. by D. Miall.

Wolman, M.G. and L.B. Leopold (1957). "River Flood Plains: Some Observations On Their Formation". In: *United States Government Printing Office* 282–C, pp. 87–107.

Zeichner, S., J. Nghiem, et al. (2020). *Models and code for Early plant organics increased global terrestrial mud deposition through enhanced flocculation*. DOI: 10.5281/zenodo.4033293.

Zeichner, S.S., J. Ngheim, et al. (2021). "Early plant organics increased global terrestrial mud deposition through enhanced flocculation". In: *Science* 371 (6528), pp. 526–529. DOI: 10.1126/science.abd0379.

Zhang, G. et al. (2013). "Effects of exopolymers on particle size distributions of suspended cohesive sediments". In: *Journal of Geophysical Research: Oceans* 118, pp. 3473–3489. DOI: 10.1002/jgrc.20263.

Supplementary Materials and Methods

Experimental Design and Construction

Two polymers, xanthan gum (a linear polysaccharide of glucose, mannose, and glucuronic acid produced by the fermentation of plant carbohydrates by the gamma-proteobacterium *Xanthomonas campestris*) and guar gum (a branched polysaccharide of galactose and mannose derived from the plant *Cyamopsis tetragonoloba*), and two clays, smectite and kaolinite, were used to quantify the effect of organic polymer on flocculation.

Natural soils are dominated by three main types of clay: kaolinite, illite and smectite (Ito and Wagai, 2017); the two clays we chose for this experiment represent mineralogical end members of dominant clay minerals in soils. Kaolinite has 1:1 tetrahedral:octahedral geometry without interlayer cations, whereas smectite has 2:1 tetrahedral:octahedral geometry with charged interlayer cations capable of undergoing distinct bonding interactions with organics.

Likewise, the two polymers chosen for this study have distinct sources, structures, branching patterns, and functional groups that affect presence and abundance within natural soils as well as bonding interactions. At circum-neutral pH, xanthan gum is a negatively charged, linear bacterial polymer, whereas guar gum is a neutral branched plant polymer capable of a wider range of molecular interactions. These molecules are long-chain polysaccharides with variable branches and charged functional groups, comparable to most dominant organic molecules synthesized by plants in their cells walls (e.g., cellulose, Popper, 2008) and common in soil and sediment environments even as microbes act to degrade and modify plant-derived organic matter (Cheshire, 1977; Wershaw, 2004). With that, we acknowledge the wide diversity of biopolymers present in natural terrestrial organic matter, but used these model polymers as representatives of the most common types of molecules, structures, and functional groups present in plant detritus and the microbes that life closely alongside them, capable of facilitating intramolecular binding interactions. Combinations of these clays and polymers captured a range of potential molecular interactions that occur in natural soils to form flocs in suspended sediment and deposit mud onto floodplains.

Humic acids offered another opportunity to test flocculation from plant-derived organic matter, typically generated via incomplete lignin degradation (Wang et al., 2017). Overall, our experiments were designed to optimize the optical quantification of settling velocity for flocs formed by model polymer-clay combinations, and are described below. Humic acid extracts have significant light absorbance at visible wavelengths, so while the results of humic acid experiments were consistent with flocculation behavior demonstrated by xanthan and guar gums, the nature of the humics interfered with the optical absorbance measurements necessary to our primary experimental setup. We applied a distinct approach to facilitate floc tracking and settling velocity calculations in humics experiments: we used kaolinite because of its high contrast and increased the total sediment concentration to 5 g/L with 2.8 weight % humic acid. Note that these sediment concentrations are at the upper range of what is typically observed in rivers (Table S6.1), and thus preclude direct comparison between humic acid results with our other experiments. However, in experiments where humic extracts were added, we observed formation of kaolinite flocs with diameters of $\sim$100 to 300 μm and settling velocities of $\sim$10^{-3} m/s; experiments with no humics added did not produce any visible flocs.

These experiments gave insight into the complex combination of factors that affect floc formation. Flocculation is driven by primary particle composition and concentration, organic composition, and turbulent shear. In our experiments, we observed that variation in any of these factors yielded measurable differences in flocculation behavior, so we designed the following conditions both to maximize reproducibility among experiments and cover concentration regimes that reflect modern river conditions (Table S6.1). For each experiment, 0.2 g of clay was weighed into a 500-mL reactor and soaked in distilled water for one (1) hour to ensure full saturation. While soaking, the polymer was weighed to distinct ratios relative to clay weight (1, 2, 5, 8 or 10 weight %) and dissolved in water using a magnetic stir bar. This range of weight percent polymer-to-clay was inspired by the methodology of previous studies of organic-driven freshwater flocculation, and aimed to capture the range of dissolved and particulate organic matter concentrations in modern rivers (Galy, Peucker-Ehrenbrink, and Eglinton, 2015; Le Quéré et al., 2018). The concentrations of dissolved and particulate organic matter in paleofluvial systems is not well-constrained beyond the organic contents measured in the resulting sedimentary deposits.

Our experimental setup consisted of a 500-mL batch reactor placed on a stir plate within a light-sensitive box constructed out of aluminum rods (Fig. 6.2A inset). A light strip was positioned behind the reactor, with a diffusive material placed in front of the light to allow equal distribution of the light throughout the flume for optimal observation of settling. The camera was positioned at a fixed distance on the other side of the settling column within the light-sensitive box to observe the full vertical range of the flume, and connected to an adjacent computer to avoid interaction with the flume or camera during the experiment. The polymer solution and additional water were added to fully fill the 500-mL reactor. The solution was stirred for thirty more minutes prior to observing settling to ensure complete suspension and homogeneous distribution of suspended sediment within the reactor. While turbulence conditions can vary for natural rivers, flow dynamics are thought to be self-similar under the same fully turbulent flow regime (Kolmogorov, 1991). We targeted the reactor to be fully turbulent (Re ~ 10^4) and set the magnetic stir bar to the identical intensity for each experiment (King, 1992).

To observe settling, stirring was stopped and photographs were taken using a Nikon D5200 at one-second intervals (Fig. 6.2A inset). Each photo was cropped to a 198- by 742-pixel matrix to ensure the suspended sediment change over time was calculated over the same interval. To perform the data analysis and characterize the experimental conditions, several control and calibration experiments were performed: First, experiments were performed without addition of organics as a control to characterize behavior of unflocculated clay (e.g., potential for abiotic aggregation). Second, measurements of light intensity were correlated to a range of clay suspended sediment concentration (0.0 g L^{-1} to 1.0 g L^{-1}, at 0.02 g L^{-1} increments), related using an exponential regression, and applied to convert photographs from experiment into concentration units. Third, to constrain the velocity of the water, a buoyant particle was captured moving within the reactor. This value was used to the Reynolds number:

$$Re = \frac{uL}{v} \tag{S6.1}$$

where u is average particle velocity (0.12 m s^{-1}), L is the diameter of the reactor (0.08 m), and v is the kinematic viscosity of the fluid (10^{-6} m^2 s^{-1}). Our flume had an approximate Reynolds number of 10048, and particles followed a chaotic trajectory confirming turbulent flow in our river analog system (King, 1992). Fourth, to confirm that we were accurately predicting settling velocities in our experimental set-up, we performed calibration experiments with 90 mesh sand (diameter = 165 μm) and compared calculated settling velocities to Stokes' settling estimates (King, 1992). Experimental and Stokes' settling velocities for fine-grained sand overlapped within uncertainty. Finally, we performed all of our experiments at least twice to quantify experimental reproducibility. Results from our full set of experiments are presented in Table S6.2 and the code for photo processing and settling velocity quantification has been made available online (Zeichner, Nghiem, et al., 2020).

Data Analysis and Modeling

Average photo pixel values were used to measure sediment concentration; changes in pixel value over the experiment were converted into sediment concentration over time using a previously determined calibration curve to match pixel intensity to concentration (Fig. S6.2). Settling velocities were calculated over the interval of maximum settling using the following relationship

$$\frac{C}{C_o} = \frac{1 - w_s}{\Delta z \times t} \tag{S6.2}$$

where C is the volumetric suspended sediment concentration of a given area in the photo matrix, C_o is the initial concentration of at that area, w_s is the average settling velocity, Δz is the change in height over the area of interest, and t is time. We fit a line using linear least squares regression between $\frac{C}{C_o}$ and time (e.g., black line in Fig. 6.2A), and calculated w_s by dividing the best-fit slope of the line by $-\Delta z$.

Equation S6.2 was derived as follows. The change in suspended concentration with respect to time, $\frac{dC}{dt}$, is related to the change in sediment concentration over the height of the cropped photo window dz by mass balance

$$\frac{dC}{dt} = \frac{d(Cw_s)}{dz} \tag{S6.3}$$

and z is the height of the window of interest, where the z direction is positive upwards. We calculated average concentration using:

$$\overline{C} = \frac{1}{\Delta z} \int_0^{\Delta z} C \, dz \tag{S6.4}$$

We substituted this relationship into equation S6.3 to get

$$\frac{d\overline{C}}{dt} = \frac{1}{\Delta z}[Cw_s|(z = \Delta z) - Cw_s|(z = 0)] \tag{S6.5}$$

The $C_w s$ term when $z = \Delta z$ is zero because the water at the top of the photo window of interest will clear, and the concentration will go to zero, so

$$\frac{d\overline{C}}{dt} = \frac{-1}{\Delta z}[Cw_s] \qquad (S6.6)$$

Based on the assumption that the settling reached its terminal settling velocity, we integrated the two sides of the equation with respect to concentration and time to find equation S6.2, using the initial concentration at t = 0 is the initial sediment concentration C_o.

Statistical analysis

We grouped our average measured settling velocities to evaluate statistically significant differences between the settling velocities observed under different experimental conditions. We formally tested the following null hypotheses H_0, with rows of Table S6.2 indicated in parentheses, to indicate which experiments were used for each test.

The addition of organics produced identical average clay settling velocities compared to clay without organics added (control: 1, 12; flocculated: 2:11, 13:29).

The addition of guar gum yielded identical average floc settling velocities compared to those produced via the addition of xanthan gum (guar: 2:6, 13:17, 23:26; xanthan 7:11, 18:22, 27:29).

The guar gum-smectite combination produced flocs with identical average settling velocities compared to those produced by any other polymer-clay combination (smectite-guar: 2:6, 23:26; everything else: 7:22, 27:29)

For each hypothesis, we tested an alternative hypothesis Ha in which the first group (i.e., "addition of organics" for hypothesis 1) produced greater settling velocities. Thus, we performed t-tests with H_0: $\mu = \mu_0$ and H_a: $\mu > \mu_0$. p-values were reported based on Welch's two-sample, one-sided t-tests using the t.test function in RStudio.

Floodplain sedimentation model

We modeled overbank flow using a one-dimensional floodplain cross-sectional profile perpendicular to a river channel, with deposition rates of sediment input spanning clay to sand sizes (Fig. 6.3A). The spatial domain we considered was half of a floodplain on one side of the channel. The model is a suspended sediment advection-settling model, similar to models derived in previous work (Lamb, McElroy, et al., 2010), and is derived in a simple form to arrive at an analytical solution.

The mass conservation of suspended sediment in one dimension is

$$\frac{\partial q_s}{\partial x} = w_s \times (e_s - C_b) \tag{S6.7}$$

in which q_s is the volumetric sediment flux per unit width ($m^2 s^{-1}$), x is the horizontal distance across the floodplain from the channel edge (m), w_s is particle settling velocity ($m s^{-1}$), e_s is the dimensionless entrainment rate from the bed, and C_b is the dimensionless near-bed volumetric sediment concentration (Parker, 1978; Parker et al., 1987). The sediment transport parameters are grain size-specific, and the generalization to many grain sizes was made in our implementation.

We assumed that sediment entrainment on the floodplain was negligible, and that sediment transport occurred due to advection without turbulent diffusion. These conditions gave $e_s = 0$ and $q_s = \overline{C}q$, respectively, where $\overline{C}$ is the depth-averaged dimensionless volumetric sediment concentration and q is the overbank water discharge per unit width m^2/s. We used $r_0 \equiv C_b/\overline{C} \geq 1$ to describe the stratification of the vertical suspended sediment concentration profile (Parker et al., 1987). Using these assumptions and definitions, we integrated Eqn. S6.7 to find

$$\overline{C}(x) = C_0 \exp \frac{-w_s r_0}{q} x \tag{S6.8}$$

in which the sediment concentration at the boundary of the channel and floodplain is C_0. We differentiated Eqn. S6.8 with respect to x and multiplied both sides by q to recover an equation for the divergence in sediment flux

$$\frac{\partial q_s}{\partial x} = -w_s r_0 C_0 \times \exp(\frac{-w_s r_0}{q} x) \tag{S6.9}$$

The divergence of the sediment flux is related to changes in floodplain elevation using the Exner conservation of sediment mass equation

$$(1 - \lambda_p)\frac{\partial \eta}{\partial t} = -\frac{\partial q_s}{\partial x} \tag{S6.10}$$

where λ_p is the porosity of the sediment deposit, η is floodplain elevation, and t is time (Paola and Voller, 2005). Substituting equation S6.9 into equation s6.10 results in an expression for the floodplain aggradation rate

$$\frac{\partial \eta_i}{\partial t} = \frac{w_{s_i} r_0 C_{0_i}}{(1 - \lambda_p)} \times exp(-\frac{w_{0_i} r_0}{q}x) \qquad (S6.11)$$

where the subscript i denotes values for the ith grain-size class when generalizing to n grain-size classes. We assumed that the parameters on the right-hand side of Eqn. S6.11 do not vary significantly with time during a characteristic flood event.

Inspection of Eqn. S6.11 revealed that flocculation leads to faster floodplain mud deposition rates, all else being equal. Across different fluvial systems, overbank discharge, sediment concentration, and grain size distribution can vary and cause different deposition rates and spatial trends in grain size composition of deposits. For any given combination of those variables, flocculation increases settling velocities $w_s i$ for mud particles. Based on Eqn S6.11, higher mud settling velocities led to two interacting effects: first, $\frac{\partial \eta_i}{\partial t}$ increases linearly with $w_s i$ and second, $\frac{\partial \eta}{\partial t}$ decays exponentially with $w_s i$ where the horizontal scale is also determined by overbank discharge q. Thus, flocculation causes overall greater mud aggradation rates, and a steeper falloff of mud deposition rate with distance from the channel.

Input model parameters for two scenarios were scaled after observed values for the Mississippi River , USA as an example of a large, low-gradient fluvial system (Jordan, 1965; Scott and Stephens, 1966), which exemplifies the deep channeled, low-gradient single-threaded rivers common before and after the Silurian Period (Ganti, Whittaker, et al., 2019). The per-width discharge q is the product of overbank flow depth and velocity. We based overbank flow depth of 2.3 m and overbank flow velocity of 1 m/s (q = 2.3 m^2/s) on direct flood measurements at the main stem Missouri River near its confluence with the Mississippi River (Wolman and Leopold, 1957). The total overbank sediment concentration sourced from the channel, which sets $C_0 = \sum C_0 i$, is independent of percent mud and normalized deposition rate, so it was set to an arbitrary constant. We specified a log-normal distribution of suspended sediment grain size at the channel boundary (median or geometric mean 5.6μm, geometric standard deviation 5.5 mm) to interpolate the total averaged concentration across 28 log-spaced size classes with minimum grain size 0.02 μm and maximum grain size 2 mm (Fig. 6.3A). For the flocculated mud scenario, all grains smaller than 20 μm were considered flocculated to a uniform floc settling velocity of 0.34

mm/s (Fig. 6.3B) following the estimated average floc settling velocity in modern rivers based on analysis of river suspended sediment concentration profiles (Lamb, McElroy, et al., 2010). Although suspended sediment may be flocculated for grains up to 40 μm (Lamb, McElroy, et al., 2010), Stokes' settling velocity predictions exceed 0.34 mm/s for grains larger than 20 μm (Ferguson and Church, 2007). Thus, a uniform settling velocity of 0.34 mm/s was only assigned to those grain size classes smaller than 20 μm to enforce the observation that flocculation generally increases settling velocity. Compared to the experiment results, the prescribed floc settling rate of 0.34 mm/s was similar to the estimated settling velocity in the 2 and 5% weight ratio smectite-guar treatment (0.265 and 0.437 mm/s, respectively; Fig. 6.2C), and fell in the range of experiment settling velocities. Based on prior work, we expected that the presence of silt, which we did not explicitly test in experiments, would further increase floc settling velocities (Tran and Strom, 2017). For coarser grains and all grains in the unflocculated case, settling velocity was calculated from reference (Ferguson and Church, 2007). We specified $r_0 = 1$ (corresponding to a uniform suspended sediment concentration profile) because the influence of r_0 was of secondary importance compared to the settling velocity.

We generated the grain size-specific floodplain model profiles according to Eqn. S6.11, and used the model profiles to assess the effect of flocculation on relative mud abundance compared to sand across a floodplain (Fig. 6.3C). We classified and aggregated these results into mud (D < 62.5 μm) and sand (D > 62.5 μm) size classes to calculate percent mud of deposits (Fig. 6.3C). Fig. 6.3C shows the proximal floodplain over a width of twice the advection length of sand particles ($\sim$170 m) to maintain appreciable sand in the domain for comparison with mud deposition (Fig. 6.3C). The advection length is the characteristic length over which a particle of a given size is transported before it settles to the bed in the absence of entrainment, and can be calculated from the per-width water discharge q divided by the settling velocity w_s of the median sand grain size (Lamb, McElroy, et al., 2010). This formulation illustrated that q controls the horizontal spatial scale of deposition, and that changing the value of q directly changed the scale of floodplain length at a fixed channel position but did not change the relative distribution of mud and sand. Beyond twice the advection length, the prevalence of mud over sand became more associated with the inability of the flow to transport sand grains and less due to the effects of flocculation. In other words, percent mud tended to unity with distance from the channel for all cases for large transport distances (many times the advection length of sand) simply because the sand fraction settled out completely.

Figure 6.3D shows the predictions of Eqn. S6.11 over a much greater distance across the floodplain. We used Eqn. S6.11 to compare relative magnitudes of mud deposition rates across a floodplain for flocculated and unflocculated scenarios (Fig. 6.3D). We summed deposition rates for mud size classes to compute the total mud deposition rate. We normalized the deposition rates by the maximum deposition rate in the flocculated case (occurring at the channel bank, $x = 0$) to better highlight the comparison.

In order to generalize the model results to fluvial settings beyond that considered here, we varied the input parameters independently to demonstrate their effects on the results across ranges of plausible values (Fig. S6.4). Specifically, the input parameters of per-width overbank discharge q, grain size distribution of suspended sediment advected from the channel, and floc settling velocity w_s, floc determine the two model results, percent mud and normalized mud deposition rate (Fig. 6.3C and 3D). We characterized variations in grain size distribution as variations in the median of the distribution D_{50} and held the standard deviation constant. While the lowland alluvial setting considered in Fig. 6.3 is likely to deposit substantial mud and preserve mudrock in the rock record, other types of river systems with different planform geometries and/or gradients can yield different model results because of correlated trends with characteristic overbank discharges, suspended sediment grain size distribution and floc settling velocity. Using the flocculated scenario inputs and results from Fig. S6.3 as a base case, we calculated percent mud and normalized mud deposition rate for each parameter of interest (q, D_{50}, and w_s, floc) by varying that parameter across realistic natural values and setting all other parameters to the same values used in the flocculated base case. We then compared each modeled scenario to an equivalent unflocculated scenario by setting sediment settling velocities according to predictions (Ferguson and Church, 2007). As a result, these tests demonstrated the likely range of model outcomes that can be associated with variations in physical properties of rivers.

The model sensitivity results for all variables demonstrated that individual variations in these parameters always produce sand-mud transitions closer to the channel compared to unflocculated scenarios (Fig. S6.4). This result was consistent with the generation of muddier floodplains caused by mud flocculation. Similarly, mud deposition rate results showed that, in most cases, mud deposition rate in a distal location from the channel can be increased by up to tenfold compared to unflocculated cases, again consistent with muddier floodplains. However, deposition rate of flocculated mud can be smaller than that of unflocculated mud for sufficiently small q < 1 m^2/s. This behavior highlighted the role of q in setting the horizontal mud advection scale. If discharge is too small, the flocculated mud will settle out completely closer to the channel such that more distal parts of the floodplain will be starved of mud. Additionally, model results for large D_{50} converged to the unflocculated case, showing that the effect of flocculation is minimal with low mud supply from the channel.

If we considered for model purposes that different fluvial systems are characterized by differences in their overbank discharge and sediment grain size distribution and that flocculation simply sets the grain size-settling velocity relationship, then the introduction of mud flocculation always produced a model response for any given river tending to muddier floodplains for realistic ranges of w_s, floc (Fig. S6.4). Compared to unflocculated results at identical discharge and grain size distribution, w_s, floc promoted higher distal floodplain mud deposition rates (up to 3.6 times greater than that of the unflocculated case) and more proximal transitions to dominantly mud deposits (up to 2 orders of magnitude closer to the channel). Thus, the role of mud flocculation alone is expected to cause more channel-proximal floodplain mud distribution and greater floodplain mud abundance, and in turn overall muddier floodplains.

Attribute	Observed condition
Suspended sed. concentrations	0.14–3.08 g/L, avg of 1.25 g/L (Dallmann et al., 2004)
Most abundant clay minerals	Illite, kaolinite, smectite, vermiculite (Ito and Wagai, 2017)
Organic concentrations	1-6% (19); 1-20%(Zhang et al., 2013)

Table S6.1: Typical natural river conditions.

Relevant characteristics of global river sediment (sed.) that guided experimental design for the flocculation experiments. Continents vary in clay mineralogy and percentage due to soil type, topography, and bedrock geology. The ranges for percent organics were based on two studies examining the role of organics in forming flocs within fluvial systems.

Clay	Polymer	E/C (%)	n	(abs(w_s), m/s)	RSE	$log_{10}(abs(w_s))$
Smectite	n/a	0	9	1.69E-06	1.57E+00	-5.77
Smectite	guar	1	5	6.24E-05	4.06E-01	-4.20
Smectite	guar	2	4	2.65E-04	5.21E-01	-3.58
Smectite	guar	5	11	4.37E-04	1.96E-01	-3.36
Smectite	guar	8	3	6.28E-04	2.43E-01	-3.20
Smectite	guar	10	4	1.28E-03	2.48E-01	-2.89
Smectite	xanthan	1	3	1.55E-06	-2.77E+01	-5.81
Smectite	xanthan	2	3	2.40E-06	-1.74E+00	-5.62
Smectite	xanthan	5	2	1.24E-05	6.75E-01	-4.91
Smectite	xanthan	8	3	3.37E-06	3.22E-02	-5.47
Smectite	xanthan	10	3	3.19E-06	-7.49E-01	-5.50
Kaolinite	n/a	0	3	2.72E-06	6.36E-07	-5.57
Kaolinite	guar	1	4	1.54E-05	2.47E-01	-4.81
Kaolinite	guar	2	3	2.64E-05	7.36E-01	-4.58
Kaolinite	guar	5	2	6.20E-06	2.90E+00	-5.21
Kaolinite	guar	8	2	3.99E-06	-2.84E-01	-5.40
Kaolinite	guar	10	2	3.32E-06	-2.96E-01	-5.48
Kaolinite	xanthan	1	2	1.82E-05	-1.19E-01	-4.74
Kaolinite	xanthan	2	2	7.44E-06	3.29E-01	-5.13
Kaolinite	xanthan	5	2	1.33E-05	-5.70E-01	-4.88
Kaolinite	xanthan	8	1	6.93E-06	n/a	-5.16
Kaolinite	xanthan	10	2	8.59E-06	-1.78E-01	-5.07
Kaolinite	humic	2.8	2	1E-03	2E-04	-3
Smectite : Kaolinite (0.25)	guar	10	1	8.72E-05	n/a	-4.06
Smectite : Kaolinite (0.50)	guar	5	1	6.34E-04	n/a	-3.20
Smectite : Kaolinite (0.50)	guar	10	1	4.63E-04	n/a	-3.33
Smectite : Kaolinite (0.75)	guar	10	1	6.44E-04	n/a	-3.19
Smectite : Kaolinite (0.50)	xanthan	2	1	3.17E-06	n/a	-5.50
Smectite : Kaolinite (0.50)	xanthan	5	1	3.61E-05	n/a	-4.44
Smectite : Kaolinite (0.50)	xanthan	10	1	8.59E-06	n/a	-5.07
90 Mesh sand	n/a	n/a	3	1.72E-02	3.54E-01	-1.76
Fine silt (theoretical)	na	na	na	1.44E-06	n/a	-5.84
Coarse silt (theoretical)	n/a	n/a	n/a	1.44E-04	n/a	-3.84
Fine Sand (theoretical)	n/a	n/a	n/a	2.29E-02	n/a	-1.64

Table S6.2: Experimental results.

Settling velocities measured in flocs experiments. Italicized rows represent settling velocities that were not meaningfully different from the clay control when accounting for the uncertainty in the sediment concentration measurements. The asterisk denotes distinct experimental setup used for humic acid experiment. RSE = relative standard error. w_s = settling velocity.

Fig. S6.1: Flocculated and unflocculated clays.

Example of the substantial impact that organic polymers can have on flocculation and settling. Unflocculated smectite (2 g/500 mL) in optical settling column after sitting overnight (left) versus flocculated, settled smectite (2 g/500 mL + 1 weight % guar gum) after t = 1 minute (right).

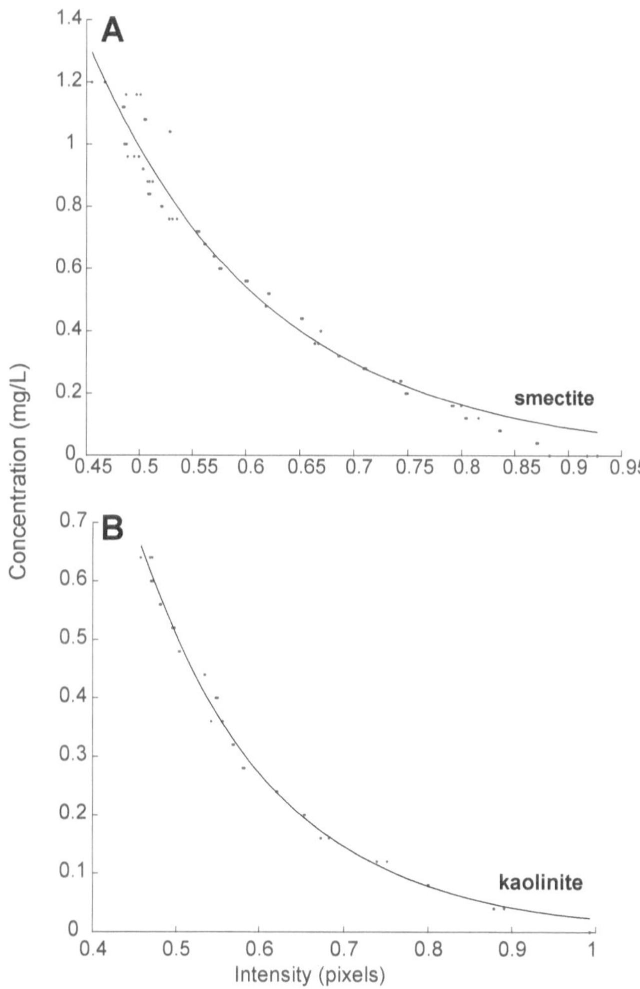

Fig. S6.2: Experimental intensity to sediment concentration calibration curves.

Calibration curves for (A) smectite and (B) kaolinite that were used to calculate light absorbance (measured by photo pixel intensity) into sediment concentration.

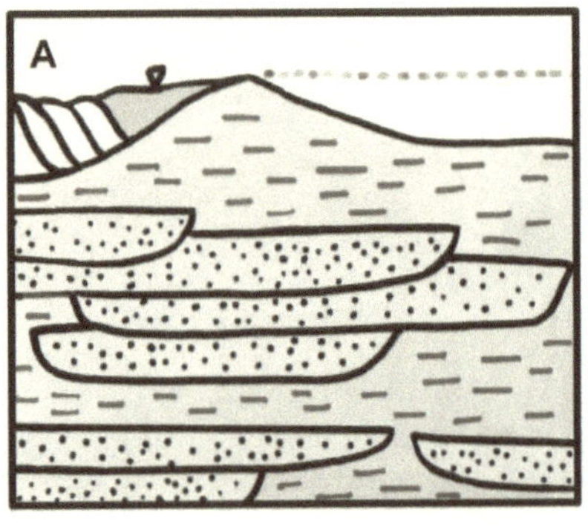

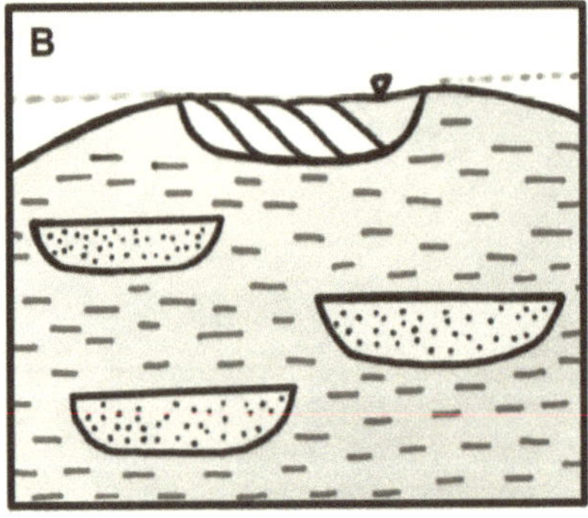

Fig. S6.3: Sand- versus mud-dominant alluvial deposits.

Schematic showing the potential distribution of mud and sand within fluvial deposits (A) without and (B) with flocculation, driven by different amounts lateral channel migration relative to avulsion. Without flocculation, banks are sandier and river lateral migration rates are faster, resulting in wider sand bodies that are more interconnected. With flocculation, banks are stronger and channel bodies are narrower, and reduced lateral migration relative to avulsion results in isolated sandstone bodies.

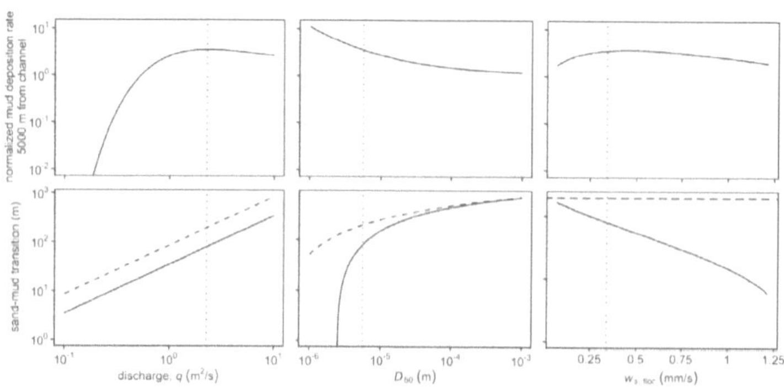

Fig. S6.4: Floodplain sediment transport model sensitivity analysis.

Model results showing the effect of overbank discharge per unit width, q, median grain size, D_{50}, and floc settling velocity w_s, floc (range of 16th to 84th percent quantiles of w_s, floc based on modern river data analyzed in Lamb, Leeuw, et al., 2020) on the main results shown in Fig. 6.3. Top row: Mud deposition rate on the floodplain at 5000 m from the channel for the flocculated case, normalized by mud deposition rate at 5000 m for the unflocculated case. Bottom row: Distance from the channel of the sand-mud transition (defined by 50% mud). The dashed curves in the bottom row panels mark the distance of the sand-mud transition in the unflocculated case calculated with Stokes' settling velocities. For each column, all other variables in the model were held constant and set to the values used in Fig. 6.3 (q = 2.3 m^2/s, D_{50} = 5.6 μm, w_s, floc = 0.34 mm/s, shown as vertical dotted lines in their respective columns).

GEOLOGISTS AND THE EARTH: BUILDING A BETTER SYMBIOSIS

Originally published as a Viewpoints article in Caltech Letters, January 5, 2021

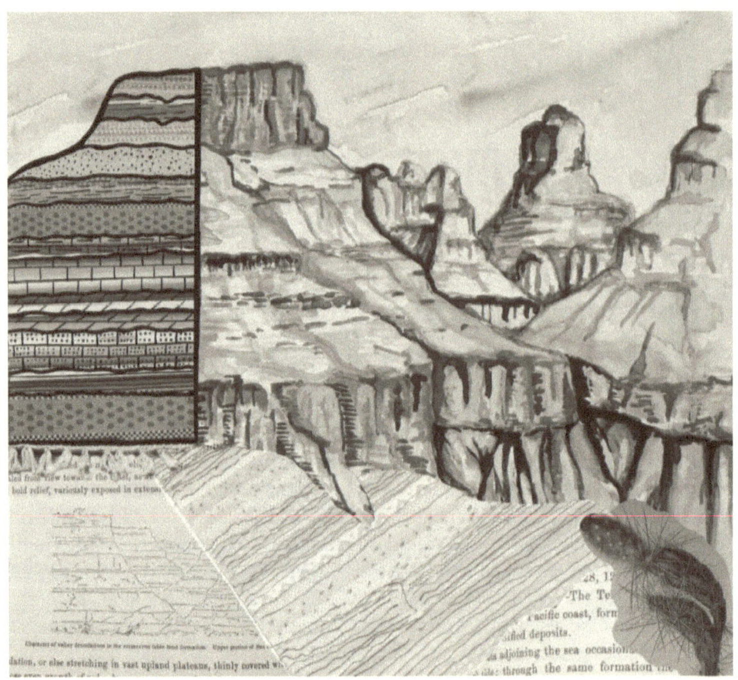

Figure A.1: *Strata.*

Mixed media collage by S. Zeichner for *Caltech Letters.*

The drone shattered the afternoon silence as it rose whirring into the air, capturing a panoramic view of our field site in the Anza Borrego desert, less than one hundred miles from the US-Mexico border. After forty minutes of troubleshooting in the sweltering heat, the sound of a functional drone brought feelings of relief. Closer to the border, drones play a much different role and evoke a very different feeling. The tension between these potential drone-provoked feelings represents one of many examples we were trying to explore on this trip.

I am no stranger to dichotomy. I am a geologist, specializing in geochemistry, and an artist; these two distinct aspects of my identity shape the way I see the world and also enabled my participation in this project, *Incendiary Traces*, led by Professor Hillary Mushkin at Caltech. For this part of the project, a group of geologists and artists traveled to the desert to explore artistic versus scientific methods of observing and recording landscapes.

Figure A.2: Surveying and surveillance.

Cecilia Sanders, a geologist and artist on the Incendiary Traces field trip, flies the drone.

Art and science both share the ultimate goal of storytelling, whether it be of a landscape, a civilization, or a natural process. Yet, science is typically seen as an "objective" perspective on natural processes, while art chooses to engage more

Figure A.3: Artists and geologists hiking during Incendiary Traces.

overtly with the complex dynamics among nature, society, and politics. Scientists frame observations of the natural world within known scientific paradigms and principles. For instance, geologists use "cross-cutting" relationships to interpret the connection between rocks in an "objective" way: younger rocks get laid down on top of, or penetrate into (in the case of molten magma) older rocks. Understanding the transitions in these strata, or rock layers, allow us to reconstruct Earth's history over billions of years.

The biological and geological worlds are closely linked, but biology defines relationships in a different way. For instance, models of biological interaction describe symbioses between individuals living together in a community. These symbioses can be beneficial for both parties (mutualistic), can neither help nor hurt either party (commensal), or can benefit one party but harm the other (parasitic).

To understand geological or biological relationships, scientists must travel around the world—often into extreme environments—to collect data. The results of these studies typically focus on the scientist's perspective exclusively, rather than engaging with the political, social, and environmental contexts of the world they're studying. While doing field work, scientists inevitably affect the places where they

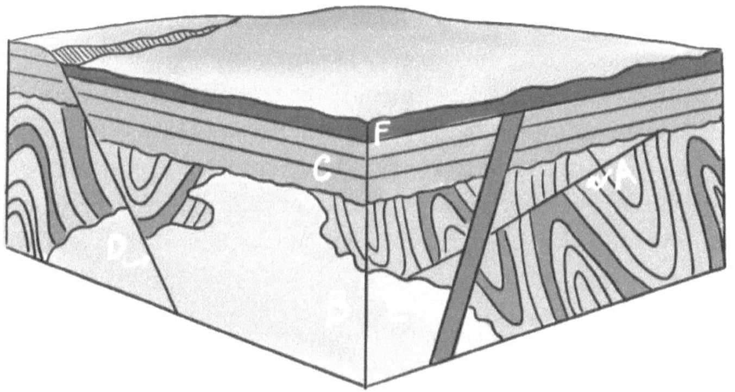

Figure A.4: Cross cutting relationships.

A cartoon of cross cutting relationships between rocks, as taught in introductory geology classes. Original sedimentary layers (light brown and grey) are first laid down in flat layers, then transformed over millions of years. A fault (A) cross cuts those layers, and then they are intruded (B) by molten rock that slowly cools. More sediment is deposited on top (C), followed by another fault (D) and intrusion (E). Each of these layers is referred to as a "stratum," and a series of layers is described as a stratigraphic column. Modern sediment will be deposited on ancient strata (F). Adapted from Wikipedia.com.

are working. Most literally, geologists consume things like food and water and collect valuable natural samples. Though this consumption is theoretically beneficial for everyone, it has the potential to harm the natural environment or people living near the sites. Usha Lingappa, a geobiologist studying the origin of photosynthesis and an artist on the *Incendiary Traces* trip, expressed concern about the interactions between geologists and their field sites by drawing a parallel to the models of biological interaction: "at best, most field geologists are neutral in their presence. At worst, I think we can verge on parasitic."

In contrast, art works to challenge the viewers' perception of the world around them, what they know about it, and how they understand their role within it. It acknowledges the impossibility of objectivity and the inherent role of both the artist and the audience in the work. These distinctions made it obvious to me, even before the trip began, that scientists and artists would ask distinct questions of their field sites. Indeed, the entire purpose of the *Incendiary Traces* project was to tease apart these dynamics in the Anza Borrego desert. Our site in Anza Borrego has a rich history of documented geologic formations and native plants. But being so close to the US-Mexico border, this rich natural history co-mingles with a tangled socio-political history. The original geologic survey of the *Incendiary Traces* field site, the US Mexico Boundary Report, was generated between 1857 and 1859. The goal of the original report was to use systematic techniques of observation and measurement to produce an objective record of the "natural properties"—plants, rocks, animals, and native populations—and potential resource value of the newly ceded territory for the US government.

One hundred fifty years later, the *Incendiary Traces* artists produced different images of the same landscape, with goals of including subjectivity, emotion, and interpretation in the work. For instance, along with hand-drawn images similar to those in the boundary report, drone footage enabled a new perspective of the landscape. Geologists often employ instruments like drones for scientific surveying, but this technique is politically charged and can evoke thoughts of surveillance for people living nearby whose stability and citizenship are in flux. Science can potentially be parasitic by taking only what is valuable to the scientific question without acknowledgment of the broader context within which the science is conducted.

Acknowledging this tension among the natural, social, and political thus poses key questions for scientists. Are scientists exempt from the broader socio-political context of their work, if the goal is to limit scientific research to observation of the natural world? What effect do we, as scientists, have on field sites, and what are the deeper social and political implications of this effect? Can consideration of additional complexity improve "objective" natural studies? In other words, how can scientists improve their research and presence in the world by considering their science through an artistic lens?

Science is connected to the rest of the world and thus is still nested in its complexities. For instance, science requires funding, which inherently situates it within a broader social and political context. As a geochemist, my research characterizes chemical patterns in carbon-based molecules that have been found on meteorites and in ancient rocks to try to understand their formation prior to the origin of life. The petroleum industry relies on a similar understanding of biomolecules, as oil and gas originate from deposits of ancient biomass. My own research motivations exist beyond those of the petroleum industry, but my work itself is not immune to

Figure A.5: Incendiary Traces artists produced drastically different interpretations of the same landscape.

its influence, and relies on its past research discoveries. Naming funding sources within published work is required, and also a step towards acknowledging potentially subjective opinions, influences or agendas behind the "objective" work.

Context gets more complex as land and natural resources become involved. Field sites are often on Native lands, and samples critical for scientific discovery may be perceived as valuable by other parties, perhaps for cultural or spiritual reasons. Regulations are set up in many places to control the use of geologic and scientific resources and avoid overt parasitism. National parks require permits to take rock samples, and international laws have been set up to regulate specific types of sampling. For instance, the Nagoya Protocol dictates how people can sample biological resources in an effort to regulate and preserve access to them (1).

In many ways, permits and regulations acknowledge the complexity and nuances of the political and cultural world that encompasses natural science, preventing exploitation of natural resources that are simultaneously seen as valuable in distinct ways by various parties. A key issue is that scientists typically do not view their field work as potentially harmful. "Geologists don't perceive the way they use natural resources as exploitative, and they see the permit process designed to target other people, like those extracting mineral resources or land disturbance," said Ted Present, a sedimentary geologist at Caltech working on reconstructing the conditions of past Earth. He advocates that, beyond being a legal requirement, permits could transform a potentially parasitic relationship into a mutualistic one. "I have come to realize that permits are actually really valuable for us as scientists. The people handling them have valuable local information. If you want to find the one good sample of a specific type of rock, they can tell you exactly where to find it" (1).

Studying the effects of humanity on the natural world is very much in the spirit of geology; scientists have the potential to benefit the socio-political context within which they are acting, and vice versa. Indeed, human development facilitates geologic research. Road creation and drone technology have allowed broad surveying and documentation of field sites, allowing access to new views of rock strata, or layers. Some areas of geology (e.g., Arctic climate change research) already have begun to incorporate Native records into their research (2), as First Peoples have long-lasting and distinct information about how the land has changed over time. In other facets of geological research, including seismological or hazard surveys, the study goals are intrinsically tied to helping communities living near the field sites become more cognizant of the potential dangers of the areas they live in.

Through my own work and the anecdotes of others, I have seen how scientists affect the environments they work in—sometimes positively, sometimes negatively. The world is constantly shifting; as scientists who study the world, we must consider how we can engage with it in a neutral or, ideally, mutualistic way. At a minimum, following government protocols and permit regulations requires us to engage with, and be mindful of, the social and political dynamics at play within our research environments. Explicit land acknowledgements or inclusion of native perspectives within published work offers another way to note the complex relationships among people, sites, and samples, although this is a relatively recent effort within the field of geology. As someone early in my career (and perhaps more optimistic than some of the more cynical, tenured generations of scientists), I believe we can choose to do more. In other fields, paradigms like bioethics have emerged as structured ways to approach these issues; in contrast, geoethics is newer and less-well known and its underlying principles are still being discussed and developed (3).

Scientists think critically about our research and our data—it is time to turn that critical eye on ourselves. We can challenge the norm of science existing in a vacuum, consider the effects of our science on people living in field areas, and explore how we can include them in the work that we do. We can set new precedents and standards for our own research and our collaborations and work to perform future work within a "geo-ethical" context (3).

Work produced by this project has been displayed within art museums across the country, including at the Museum of Contemporary Art in Tuscon, AZ, and the Pasadena Armory for the Arts in Pasadena, CA.

References:

1: "About the Nagoya Protocol," Convention on Biological Diversity.

2: Couzin, J. 2007. "Opening Doors to Native Knowledge." Science.

3: Geoethics.